国家林业和草原局普通高等教育"十三五"规划教材

FOREST PLANTS
森林植物

谢春平　南程慧　主编

中国林业出版社

图书在版编目（CIP）数据

森林植物=FOREST PLANTS：英文／谢春平，南程慧主编. —北京：中国林业出版社，2021.12
国家林业和草原局普通高等教育"十三五"规划教材
ISBN 978-7-5219-1510-5

Ⅰ.①森… Ⅱ.①谢… ②南… Ⅲ.①森林植物-高等学校-教材-英文 Ⅳ.①S718.3

中国版本图书馆 CIP 数据核字（2022）第 007577 号

中国林业出版社教育分社

策划编辑：肖基浒　杨长峰　　　责任编辑：肖基浒
电　　话：（010）83143555　　　传　　真：（010）83143516

出版发行　中国林业出版社（100009　北京市西城区德内大街刘海胡同7号）
　　　　　E-mail:jiaocaipublic@163.com　电话：（010）83143500
　　　　　http://www.forestry.gov.cn/lycb.html
印　刷　北京中科印刷有限公司
版　次　2021年12月第1版
印　次　2021年12月第1次印刷
开　本　787mm×1092mm　1/16
印　张　18.75
字　数　560千字
定　价　58.00元

未经许可，不得以任何方式复制或抄袭本书之部分或全部内容。

版权所有　侵权必究

Foreword

China has emerged as a global pioneer in the development of ecological society. The demands for environmental security are rising as China's economy develops and opens up to the rest of the world. Externally, China actively participates in the global ecological protection work. Internally, China is actively carrying out the construction of beautiful China, protect the green water and green mountains, and realize the sustainable development of the Chinese nation. Therefore, modern university students should not only have a solid basic knowledge of forest plant, but also have the corresponding bilingual ability and a broad international perspective. Forest Plants is a basic course not only for most forestry colleges and universities in China, but also for ecological and environmental majors in other general types of institutions. The editors have been conducting bilingual teaching of this course for several years. Based on the revision of previous years' teaching handouts, the publication of a set of bilingual "Forest Plants" textbook suitable for the teaching needs of China's forestry colleges and universities came into being.

This textbook includes five parts: introduction, morphological features of forest plant, fundamentals of plant taxonomy, gymnosperms and angiosperms. (1) The introduction mainly introduces the basic knowledge related to forest plants, as well as the current status of Chinese forest resources, plant diversity, botany, and the purpose and methods of this course. (2) Morphological features of forest plant is described in terms of roots, stems, leaves, flowers and fruits, and description terms of common plants are displayed with Chinese annotations. (3) Fundamentals of plant taxonomy introduces the botanical nomenclature, phylogenetic systems of classification, and taxonomy keys. (4) Gymnosperms are arranged in the Zheng's classification system, introducing common 8 families and 28 species in China. (5) The angiosperms are arranged in the Hutchinson classification system, introducing 57 families with nearly 200 species commonly found in Eastern China. A total of 223 common forest tree species have been selected in the textbook, and each species is illustrated with a beautiful color picture. While introducing each species, a brief introduction to similar or closely related species is provided for students to compare.

The selected families and tree species in the textbook are based on the principle of "emphasis on the foundation", so that students can master basic identification skills of forest plant as much as possible, without emphasizing over broad species. Secondly, some rare and endangered species is appropriately selected (the species names are followed by Ⅰ and Ⅱ to indicate Class Ⅰ and Ⅱ, respectively) to meet the teaching needs of some institutions.

This textbook was selected as one of the "Thirteenth Five-Year Plan" key textbook for general higher education of National Forestry and Grassland Administration. At the same time, it received funding from the Teaching Material Construction Project of Nanjing Forest Police College and the Jiangsu Province Key Discipline Project (2016-2020) (Forestry) of the Thirteenth Five-Year Plan of Jiangsu Province Education Department, respectively.

The textbook is edited chiefly by Xie Chunping (谢春平) and Nan Chenghui (南程慧), with the participation of Wu Xiankun (吴显坤), Xue Xiaoming (薛晓明), Fang Yan (方彦), and Chen Yunxia (陈云霞). Xie Chunping is responsible for the unification and finalization of the textbook. The pictures in the textbook were provided by Nan Chenghui and Xie Chunping, and the ink drawings of plant morphology were drawn by Xie Chunping. Yan Yi (颜易), from Nanjing Forest Police College, assisted in drawing some of the ink drawings, for which we would like to express our sincere gratitude.

The textbook can be used as a reference for related majors such as agriculture, forestry, normal education, and comprehensive universities.

As one of the few bilingual textbooks of forest plant for undergraduates in China, the authors can refer to a limited number of models. However, due to the limited level of the editors, inexperience, time and other reasons, there are bound to be errors and shortcomings in the textbook, and we honestly expectpeers and readers to criticize and point out our mistakes.

Editors of Forest Plants
July, 2020

前 言

中国已成为世界生态文明建设的引领者。随着我国经济社会发展和对外开放程度的不断深入，对环境保护的要求也日益提升。对外，我国积极参与全球生态环境保护工作；对内，积极开展美丽中国建设，保护绿水青山，实现中华民族的永续发展。因此，现代大学生不仅需要具有扎实的森林植物学基础知识，更应具有相应的双语运用能力和宽广的国际视野。《森林植物》不仅是我国多数林业高校的基础课程，也是其他普通高等院校生态环境类专业的基础课程。编者已开展本课程的双语教学数年，在历年教学讲义修订的基础上，整理出版这一套适合我国林业高校教学需求的双语《森林植物》教材。

本教材主要包括了绪论、森林植物形态基础、植物分类学基础、裸子植物各论和被子植物各论5部分内容。(1)绪论部分主要介绍了与森林植物相关的基础知识，同时包括了我国森林资源概况、植物多样性、植物学组成以及本课程学习的目的与方法。(2)森林植物形态基础以根、茎、叶、花、果各部为描述对象，介绍了常见的植物描述名词，同时加以中文注释。(3)植物分类学基础介绍了植物的命名、植物分类系统和植物检索表的内容。(4)裸子植物各论以郑万钧系统为排列，介绍常见8个科28种裸子植物。(5)被子植物各论以哈钦松系统为排列，介绍华东地区常见的57个科的植物，种类近200种。全书共计选入223种常见的森林树种，每个物种均配有精美的植物彩图。在介绍各物种的同时，对相似或近缘的物种进行相应的简要介绍，供学生比较学习。

教材入选的科及树种一是以"重基础"为原则，让学生尽可能掌握基本的森林植物识别技能，而不强调过于宽泛的种类；二是适当增加了少数珍稀濒危种类的介绍(种名后附以Ⅰ、Ⅱ分别表示一级和二级)，满足部分院校教学的需求。

本教材被遴选为国家林业和草原局普通高等教育"十三五"规划教材，同时获得南京森林警察学院教务处教材建设项目与江苏省教育厅"十三五"江苏省重点学科项目(2016—2020)(林学)资助。

本书由谢春平、南程慧主编，吴显坤、薛晓明、方彦及陈云霞参编，谢春平负责对全书进行统稿和定稿工作。全书彩色图片由南程慧和谢春平提供，墨线图由谢春平绘制。南京森林警察学院颜易同学协助绘制了部分墨线图,在此表示诚挚谢意。

本书可供农林、师范、综合性大学等相关专业参考使用。

作为国内为数不多面向本科生的森林植物双语教材,作者可参考范本有限;同时也由于编者水平有限、经验不足、时间仓促等原因,书中定有错误与缺点,敬请各位同行与读者批评指正。

Forest Plants 编写组

2020 年 7 月

Contents

Foreword

Chapter 0　Introduction .... (1)
　0.1　Related concepts of forest .... (2)
　　　0.1.1　Plant, forest and forest plants .... (2)
　　　0.1.2　Wild plants .... (2)
　　　0.1.3　National key protected wild plants .... (3)
　　　0.1.4　Rare plants .... (3)
　　　0.1.5　CITES .... (4)
　　　0.1.6　Old and notable trees .... (5)
　0.2　Forest resources and current status in China .... (5)
　　　0.2.1　Forest resources .... (5)
　　　0.2.2　Forest vegetation types .... (6)
　　　0.2.3　Plant diversity in China .... (8)
　　　0.2.4　Current status of endangered species in China .... (9)
　0.3　Legislation conservation of plant diversity in China .... (10)
　0.4　Botany subdisciplines .... (11)
　0.5　Aims for the course .... (13)

Chapter 1　Morphological Features of Forest Plants .... (15)
　1.1　Root .... (15)
　1.2　Stem .... (16)
　　　1.2.1　Habit .... (16)
　　　1.2.2　Twig .... (16)
　　　1.2.3　Branching type .... (18)
　　　1.2.4　Modification of stem .... (19)
　1.3　Leaf .... (20)
　　　1.3.1　Structure .... (20)

  1.3.2 Morphology ………………………………………………………… (20)

  1.3.3 Modification of leaf ……………………………………………… (24)

1.4 Flower ……………………………………………………………………………… (25)

  1.4.1 Structure …………………………………………………………………… (25)

  1.4.2 Types of flower ………………………………………………………… (26)

  1.4.3 Morphology of main parts ……………………………………… (26)

  1.4.4 Inflorescence …………………………………………………………… (29)

1.5 Fruit ………………………………………………………………………………… (31)

  1.5.1 Structure …………………………………………………………………… (31)

  1.5.2 Types ………………………………………………………………………… (31)

Chapter 2 Foundamentals of Plant Taxonomy ……………… (35)

2.1 Botanical nomenclature ……………………………………………… (35)

  2.1.1 Plant taxonomy hierarchy ……………………………………… (36)

  2.1.2 Scientific names (Binomial) ………………………………… (37)

2.2 Phylogenetic system of classification …………………………… (37)

  2.2.1 Engler system ………………………………………………………… (38)

  2.2.2 Hutchinson system ………………………………………………… (39)

  2.2.3 Cronquist system …………………………………………………… (40)

  2.2.4 Angiosperm Phylogeny Group (APG) …………………… (40)

2.3 Taxonomic keys ………………………………………………………………… (43)

  2.3.1 Types ………………………………………………………………………… (43)

  2.3.2 Use and construct a taxonomic keys ……………………… (44)

Chapter 3 Gymnosperms 裸子植物 ………………………………… (47)

3.1 Cycadaceae 苏铁科 ……………………………………………………… (48)

 Cycas Linn. 苏铁属 Ⅰ ………………………………………………………… (48)

3.2 Ginkgoaceae 银杏科 ……………………………………………………… (49)

 Ginkgo Linn. 银杏属 …………………………………………………………… (49)

3.3 Pinaceae 松科 ………………………………………………………………… (50)

 1. *Keteleeria* Carr. 油杉属 ……………………………………………… (50)

 2. *Abies* Mill. 冷杉属 ………………………………………………………… (51)

 3. *Pseudotsuga* Carr. 黄杉属 Ⅱ …………………………………… (52)

 4. *Tsuga* (Endl.) Carr. 铁杉属 ……………………………………… (53)

 5. *Cathaya* Chun et Kuang 银杉属 ……………………………… (54)

 6. *Picea* A. Dietr. 云杉属 ………………………………………………… (55)

 7. *Larix* Mill. 落叶松属 …………………………………………………… (56)

		8. *Pseudolarix* Gord. 金钱松属	(57)
		9. *Cedrus* Trew 雪松属	(58)
		10. *Pinus* Linn. 松属	(59)
3.4	**Taxodiaceae 杉科**		(62)
		1. *Cunninghamia* R. Brown ex Rich. et A. Rich. 杉木属	(63)
		2. *Cryptomeria* D. Don 柳杉属	(64)
		3. *Taxodium* Rich. 落羽杉属	(65)
		4. *Metasequoia* Miki ex Hu et Cheng 水杉属	(66)
3.5	**Cupressaceae 柏科**		(67)
		1. *Platycladus* Spach 侧柏属	(67)
		2. *Cupressus* Linn. 柏木属	(68)
		3. *Chamaecyparis* Spach 扁柏属	(69)
		4. *Fokienia* A. Henry et H. H. Thomas 福建柏属	(70)
		5. *Juniperus* Linn. 刺柏属	(71)
3.6	**Podocarpaceae 罗汉松科**		(73)
	Podocarpus L. Her. ex Persoon 罗汉松属 II		(73)
3.7	**Cephalotaxaceae 三尖杉科**		(74)
	Cephalotaxus Sieb. et Zucc. ex Endl. 三尖杉属		(74)
3.8	**Taxaceae 红豆杉科**		(75)
		1. *Taxus* Linn. 红豆杉属 I	(75)
		2. *Torreya* Arn. 榧树属 II	(77)

Chapter 4   Angiosperms 被子植物 .... (79)

4.1	**Magnoliaceae 木兰科**	(80)
	1. *Magnolia* Linn. 木兰属	(80)
	2. *Michelia* Linn. 含笑属	(86)
	3. *Liriodendron* Linn. 鹅掌楸属	(89)
4.2	**Illiciaceae 八角科**	(90)
	Illicium Linn. 八角属	(91)
4.3	**Lauraceae 樟科**	(92)
	1. *Cinnamomum* Schaeff. 樟属	(92)
	2. *Phoebe* Nees 楠属	(94)
	3. *Machilus* Nees 润楠属	(98)
	4. *Sassafras* J. Presl 檫木属	(100)
	5. *Litsea* Lam. 木姜子属	(101)
	6. *Lindera* Thunb. 山胡椒属	(102)

4.4 Rosaceae 蔷薇科 (104)

Ⅰ. Spiraeoideae 绣线菊亚科 (104)
1. *Spiraea* Linn. 绣线菊属 (104)
2. *Exochorda* Lindl. 白鹃梅属 (105)

Ⅱ. Maloideae 苹果亚科 (107)
3. *Pyracantha* M. Roem. 火棘属 (107)
4. *Crataegus* Linn. 山楂属 (108)
5. *Photinia* Lindl. 石楠属 (109)
6. *Eriobotrya* Lindl. 枇杷属 (111)
7. *Sorbus* Linn. 花楸属 (112)
8. *Pyrus* Linn. 梨属 (114)
9. *Malus* Mill. 苹果属 (116)
10. *Chaenomeles* Lindl. 木瓜属 (119)

Ⅲ. Rosoideae 蔷薇亚科 (120)
11. *Rosa* Linn. 蔷薇属 (121)
12. *Kerria* Candolle 棣棠属 (124)
13. *Rubus* Linn. 悬钩子属 (124)

Ⅳ. Prunoideae 李亚科 (126)
14. *Amygdalus* Linn. 桃属 (126)
15. *Armeniaca* Scop. 杏属 (127)
16. *Prunus* Linn. 李属 (130)
17. *Cerasus* Mill. 樱属 (131)

4.5 Calycanthaceae 蜡梅科 (134)
1. *Chimonanthus* Lindl. 蜡梅属 (134)
2. *Calycanthus* Linn. 夏蜡梅属 (135)

4.6 Mimosaceae 含羞草科 (136)
Albizia Durazz. 合欢属 (137)

4.7 Caesalpiniaceae 苏木科 (139)
1. *Caesalpinia* Linn. 云实属 (139)
2. *Gleditsia* Linn. 皂荚属 (140)
3. *Cercis* Linn. 紫荆属 (141)

4.8 Papilionaceae 蝶形花科 (142)
1. *Ormosia* Jacks. 红豆树属Ⅱ (142)
2. *Sophora* Linn. 槐属 (144)
3. *Robinia* Linn. 刺槐属 (145)

4. *Wisteria* Nutt. 紫藤属 ……………………………………………………… (146)

5. *Indigofera* Linn. 木蓝属 ……………………………………………………… (147)

6. *Dalbergia* Linn. f. 黄檀属 …………………………………………………… (148)

7. *Lespedeza* Michx. 胡枝子属 ………………………………………………… (150)

4.9 **Styracaceae 安息香科** …………………………………………………………… (151)

1. *Sinojackia* Hu 秤锤树属Ⅱ …………………………………………………… (151)

2. *Styrax* Linn. 安息香属 ……………………………………………………… (152)

4.10 **Symplocaceae 山矾科** ………………………………………………………… (153)

Symplocos Jacq. 山矾属 ……………………………………………………… (153)

4.11 **Cornaceae 山茱萸科** …………………………………………………………… (155)

Cornus Linn. 山茱萸属 ……………………………………………………… (155)

4.12 **Alangiaceae 八角枫科** ………………………………………………………… (158)

Alangium Lam. 八角枫属 …………………………………………………… (158)

4.13 **Nyssaceae 蓝果树科** …………………………………………………………… (159)

1. *Camptotheca* Dec. 喜树属 …………………………………………………… (159)

2. *Davidia* Baill. 珙桐属 ………………………………………………………… (160)

4.14 **Araliaceae 五加科** ……………………………………………………………… (162)

1. *Kalopanax* Miq. 刺楸属 ……………………………………………………… (162)

2. *Eleutherococcus* Maxim. 五加属 …………………………………………… (163)

3. *Fatsia* Dec. et Planch. 八角金盘属 ………………………………………… (164)

4.15 **Caprifoliaceae 忍冬科** ………………………………………………………… (165)

1. *Viburnum* Linn. 荚蒾属 ……………………………………………………… (165)

2. *Lonicera* Linn. 忍冬属 ……………………………………………………… (167)

4.16 **Hamamelidaceae 金缕梅科** …………………………………………………… (169)

1. *Liquidambar* Linn. 枫香树属 ………………………………………………… (169)

2. *Loropetalum* R. Brown 檵木属 ……………………………………………… (170)

3. *Parrotia* C. A. Mey. 银缕梅属 ……………………………………………… (171)

4.17 **Platanaceae 悬铃木科** ………………………………………………………… (172)

Platanus Linn. 悬铃木属 …………………………………………………… (172)

4.18 **Buxaceae 黄杨科** ……………………………………………………………… (173)

Buxus Linn. 黄杨属 ………………………………………………………… (173)

4.19 **Salicaceae 杨柳科** ……………………………………………………………… (174)

1. *Populus* Linn. 杨属 …………………………………………………………… (174)

2. *Salix* Linn. 柳属 ……………………………………………………………… (175)

4.20 **Myricaceae 杨梅科** …………………………………………………………… (176)

Myrica Linn. 杨梅属 ··· (177)

4.21 Betulaceae 桦木科 ··· (178)
    1. *Betula* Linn. 桦木属 ··· (178)
    2. *Alnus* Mill. 桤木属 ··· (179)
    3. *Corylus* Linn. 榛属 ··· (180)
    4. *Carpinus* Linn. 鹅耳枥属 ··· (181)

4.22 Fagaceae 壳斗科 ··· (183)
    1. *Fagus* Linn. 水青冈属 ··· (183)
    2. *Castanea* Mill. 栗属 ··· (184)
    3. *Castanopsis* (D. Don) Spach 栲属(锥属) ···························· (186)
    4. *Lithocarpus* Blume 柯属(石栎属) ······································· (188)
    5. *Cyclobalanopsis* Oersted 青冈属 ··· (189)
    6. *Quercus* Linn. 栎属 ·· (190)

4.23 Juglandaceae 胡桃科 ·· (194)
    1. *Juglans* Linn. 胡桃属 ··· (195)
    2. *Pterocarya* Kunth 枫杨属 ·· (196)
    3. *Cyclocarya* Iljinsk. 青钱柳属 ·· (197)
    4. *Carya* Nutt. 山核桃属 ·· (198)
    5. *Platycarya* Sieb. et Zucc. 化香树属 ···································· (200)

4.24 Ulmaceae 榆科 ··· (201)
    1. *Ulmus* Linn. 榆属 ·· (201)
    2. *Zelkova* Spach 榉属 ··· (203)
    3. *Celtis* Linn. 朴属 ··· (204)
    4. *Aphananthe* Planch. 糙叶树属 ··· (205)
    5. *Pteroceltis* Maxim. 青檀属 ·· (205)

4.25 Moraceae 桑科 ··· (206)
    1. *Maclura* Nutt. 橙桑属(柘属) ··· (207)
    2. *Morus* Linn. 桑属 ·· (208)
    3. *Broussonetia* L'Héritier ex Vent. 构属 ································· (209)
    4. *Ficus* Linn. 榕属 ··· (210)

4.26 Eucommiaceae 杜仲科 ·· (211)
    Eucommia Oliv. 杜仲属 ··· (211)

4.27 Pittosporaceae 海桐花科 ··· (212)
    Pittosporum Banks ex Gaertn. 海桐花属 ································· (212)

4.28 Tiliaceae 椴树科 ··· (213)

  1. *Tilia* Linn. 椴树属 (213)
  2. *Grewia* Linn. 扁担杆属 (215)
4.29 **Elaeocarpaceae 杜英科** (216)
  Elaeocarpus Linn. 杜英属 (216)
4.30 **Sterculiaceae 梧桐科** (217)
  1. *Firmiana* Marsili 梧桐属 (217)
  2. *Reevesia* Lindl. 梭罗树属 (217)
4.31 **Malvaceae 锦葵科** (219)
  Hibiscus Linn. 木槿属 (219)
4.32 **Theaceae 山茶科** (221)
  1. *Camellia* Linn. 山茶属 (222)
  2. *Schima* Reinw. ex Blume 木荷属 (224)
4.33 **Actinidiaceae 猕猴桃科** (225)
  Actinidia Lindl. 猕猴桃属 (225)
4.34 **Dipterocarpaceae 龙脑香科** (227)
  Parashorea Kurz 柳安属 (228)
4.35 **Ericaceae 杜鹃花科** (229)
  1. *Rhododendron* Linn. 杜鹃属 (229)
  2. *Vaccinium* Linn. 越橘属 (231)
4.36 **Hypericaceae 金丝桃科** (232)
  Hypericum Linn. 金丝桃属 (233)
4.37 **Punicaceae 石榴科** (234)
  Punica Linn. 石榴属 (234)
4.38 **Aquifoliaceae 冬青科** (235)
  Ilex Linn. 冬青属 (235)
4.39 **Celastraceae 卫矛科** (238)
  Euonymus Linn. 卫矛属 (239)
4.40 **Elaeagnaceae 胡颓子科** (242)
  Elaeagnus Linn. 胡颓子属 (242)
4.41 **Rhamnaceae 鼠李科** (243)
  1. *Hovenia* Thunb. 枳椇属 (244)
  2. *Paliurus* Mill. 马甲子属 (245)
  3. *Sageretia* Brongn. 雀梅藤属 (246)
  4. *Rhamnus* Linn. 鼠李属 (247)
4.42 **Vitaceae 葡萄科** (248)

  1. *Vitis* Linn. 葡萄属 ……………………………………………………………… (248)

  2. *Ampelopsis* Mich. 蛇葡萄属 …………………………………………………… (249)

4.43 **Ebenaceae 柿树科** ……………………………………………………………… (250)

  Diospyros Linn. 柿属 ……………………………………………………………… (250)

4.44 **Simaroubaceae 苦木科** ………………………………………………………… (252)

  Ailanthus Desf. 臭椿属 …………………………………………………………… (252)

4.45 **Meliaceae 楝科** ………………………………………………………………… (253)

  Melia Linn. 楝属 …………………………………………………………………… (253)

4.46 **Sapindaceae 无患子科** ………………………………………………………… (254)

  1. *Sapindus* Linn. 无患子属 ……………………………………………………… (254)

  2. *Koelreuteria* Laxm. 栾树属 …………………………………………………… (255)

4.47 **Anacardiaceae 漆树科** ………………………………………………………… (256)

  1. *Pistacia* Linn. 黄连木属 ………………………………………………………… (256)

  2. *Rhus* Linn. 盐肤木属 …………………………………………………………… (257)

  3. *Toxicodendron* Mill. 漆属 ……………………………………………………… (258)

4.48 **Aceraceae 槭树科** ……………………………………………………………… (259)

  Acer Linn. 槭属 …………………………………………………………………… (259)

4.49 **Hippocastanaceae 七叶树科** …………………………………………………… (262)

  Aesculus Linn. 七叶树属 ………………………………………………………… (263)

4.50 **Oleaceae 木犀科** ……………………………………………………………… (264)

  1. *Forsythia* Vahl 连翘属 ………………………………………………………… (264)

  2. *Jasminum* Linn. 素馨属 ………………………………………………………… (265)

  3. *Osmanthus* Lour. 木犀属 ……………………………………………………… (266)

  4. *Ligustrum* Linn. 女贞属 ………………………………………………………… (268)

4.51 **Apocynaceae 夹竹桃科** ………………………………………………………… (269)

  1. *Nerium* Linn. 夹竹桃属 ………………………………………………………… (269)

  2. *Trachelospermum* Lem. 络石属 ………………………………………………… (270)

4.52 **Rubiaceae 茜草科** ……………………………………………………………… (271)

  1. *Gardenia* J. Ellis 栀子属 ………………………………………………………… (271)

  2. *Emmenopterys* Oliv. 香果树属 ………………………………………………… (272)

4.53 **Bignoniaceae 紫葳科** …………………………………………………………… (273)

  Catalpa Scop. 梓属 ……………………………………………………………… (273)

4.54 **Verbenaceae 马鞭草科** ………………………………………………………… (274)

  1. *Vitex* Linn. 牡荆属 ……………………………………………………………… (274)

  2. *Clerodendrum* Linn. 大青属 …………………………………………………… (274)

  3. *Callicarpa* Linn. 紫珠属 ……………………………………………………（276）
4.55 Lythraceae 千屈菜科 ……………………………………………………（277）
  Lagerstroemia Linn. 紫薇属 ……………………………………………………（277）
4.56 Scrophulariaceae 玄参科 …………………………………………………（278）
  Paulownia Sieb. et Zucc. 泡桐属 ………………………………………………（278）
4.57 Arecaceae 棕榈科 …………………………………………………………（280）
  Trachycarpus H. Wendl. 棕榈属 ………………………………………………（280）
4.58 Poaceae 禾本科 ……………………………………………………………（281）
  Phyllostachys Sieb. et Zucc. 刚竹属 …………………………………………（281）

Chapter 0

Introduction

Forests are the most dominant and complicated terrestrial ecosystem of the earth, and are distributed around the globe. Forests account for 75% of the gross primary production of the Earth's biosphere, and contain 80% of the Earth's plant biomass. Net primary production is estimated at 21.9 gigatonnes carbon per year for tropical forests, 8.1 for temperate forests, and 2.6 for boreal forests. Forest is the material basis that the human society depends for subsistence and the cradle of nurturing the human race. Forest not only provides timber and other forest products for economic and social development and the people's life, but possesses multiple ecological functions such as water resources conservation, water and soil conservation and ecological balance maintenance, and supplies the human beings with venues for relaxation, recreation and other social services. Forest has become indispensable natural resources for the mankind.

China has a vast territory with numerous rivers, lakes and crisscross mountains. The diverse landform types and differential hydrothermal conditions in the latitudinal, longitudinal and vertical terrains have formed a complex natural and geographical environment. China is endowed by forest resources with a great variety of biological species and vegetation types, which have provided the mankind with a wealth of eco-products, wood products and eco-cultural services. In China, the forest area is around 221 million hectares and the forest coverageis 23.04% of the total land area (Ministry of Natural Resources of the PRC, 2020). The forest stock volume reaches over 17000 million cubic meters. Therefore, forest plays an important role in national economics construction and environmental conservation.

To start the course, it is necessary to understand some basic concepts and knowledge related to the course as follows.

0.1 Related concepts of forest

0.1.1 Plant, forest and forest plants

All living things were traditionally placed into one of two groups, plants and animals. This classification may date from Aristotle, who made the distinction between plants, which generally do not move, and animals, which often are mobile to catch their food. Much later, when Linnaeus created the basis of the modern system of scientific classification, these two groups became the kingdoms Vegetabilia (later Metaphyta or Plantae) and Animalia (also called Metazoa). Since then, it has become clear that the plant kingdom as originally defined included several unrelated groups, like the fungi and several groups of algae were removed to new kingdoms. However, these organisms are still often considered plants, particularly in popular contexts.

The term 'plant' generally implies the possession of the following traits: multicellularity, possession of cell walls containing cellulose, and the ability to carry out photosynthesis with primary chloroplasts. In order to define the term clearly, plant, any multicellular eukaryotic life-form, are characterized by (1) photosynthetic nutrition (a characteristic possessed by all plants except some parasitic plants and underground orchids), in which chemical energy is produced from water, minerals, and carbon dioxide with the aid of pigments and the radiant energy of the sun, (2) essentially unlimited growth at localized regions, (3) cells that contain cellulose in their walls and are therefore to some extent rigid, (4) the absence of organs of locomotion, resulting in a more or less stationary existence, (5) the absence of nervous systems, and (6) life histories that show an alteration of haploid and diploid generations, with the dominance of one over the other being taxonomically significant.

A forest is a region, which consists of several components that can be narrouly divided into two categories, biotic (living) and abiotic (non-living) components. The living parts include trees, shrubs, vines, grasses and other herbaceous (non-woody) plants, mosses, algae, fungi, insects, mammals, birds, reptiles, amphibians, and microorganisms living on the plants and animals and in the soil. Forests are essential to life on the Earth. We depend on forests for our survival, from the air we breathe to the wood we use. Besides providing habitats for animals and livelihoods for humans, forests also offer watershed protection, prevent soil erosion and mitigate climate change, without which people would not be living at the level of comfort we have today.

The plants live in forest called forest plants. There are various kinds of plant life-form in the forest, and they may be classified as trees, shrubs, herbs (forbs and graminoids), etc. In this textbook, we focus on woody plants particularly.

0.1.2 Wild plants

When it comes to plants and plant organisms, the word "wild" applies to those that emerge naturally in self-maintaining colonies in natural or semi-natural environments and may survive without the intervention of humans. The term is contrasted with "cultivated" or "domesticated" plants or plant species that have arisen through human action, such as selection or breeding, and that depend on management for their continued existence. According to *Regulations on Wild Plants Protection*: wild plants protected under these regulations refer to plants growing in natural conditions, which are precious or which are rare or near extinction and of important economic, scientific or cultural value. As regards the protection of medicinal wild plants and wild plants within urban gardens, nature reserves and scenic spots, other relevant laws and regulations shall also apply.

0.1.3 National key protected wild plants

The Chinese government has taken steps to protect wild plant life, especially species with small numbers, with such measures as *the Catalogue of the National Protected Key Wild Plants* (the First Batch); the State Council issued the catalogue in 1999 listing 246 species and all species belonging to 6 genus and 2 families that are legally protected. In the list, protected wild plants are divided into Grade I and Grade II. Those Grade I plants indicate that endangered and rare plants with important value of science, economy and culture, for instance, *Parrotia subaequalis*, *Metasequoia glyptostroboides*, *Cathaya argyrophylla*, etc. Those Grade II plants denote that endangered, vulnerable and rare plants with vital value of science and economy, for example *Pseudolarix amabilis*, *Fokienia hodginsii*, *Liriodendron chinense*, *Phoebe chekiangensis*, etc.

0.1.4 Rare plants

Rare plants may be scarce because the total population of the species may have just a few individuals, or be restricted to a narrow geographic range, or both. Some rare plants occur sparsely over a broad area. Other rare plants have many individuals, but these are crowded into a tiny area with special environment; in some cases, a single county or canyon. A third kind of rare plants are those with both few individuals and a narrow geographic range: these are the very rarest plants. Therefore, they generally have characteristics with narrow geographic range, few individuals or within special environment.

We are curious as to why certain plants are so scarce. Some plants are naturally rare with mystery reasons. These rare plants are not necessarily in danger of extinction. If their habitat is secure and they continue to reproduce in the wild, no intervention is needed. Botanists do agree that rare plants are more likely to become extinct than more common species. There are particular life history characteristics recognized by scientists as increasing a plant species' risk of extinction. Botanists also notice that other plants that were formerly more common have become rare because of changes in their environment. These changes are often brought on directly or indirectly by people's patterns of settlement, transportation, recreation, and use of natural resources. We can help rare species recover and even thrive sometimes by making changes in our own behaviors. For most species, rarity results from some combination of anthropogenic (human-induced) and evolutionary ('natural') factors rather than a single cause.

Generally, the IUCN (International Union for Conservation of Nature) Red List Categories and Criteria are an easily and widely understood system for classifying species at high risk of global extinction. The IUCN Red List is a critical indicator of the health of the world's biodiversity. Far more than a list of species and their status, it is a powerful tool to inform and catalyse action for biodiversity conservation and policy change, critical to protecting the natural resources we need to survive. It provides information about range, population size, habitat and ecology, use and/or trade, threats, and conservation actions that will help inform necessary conservation decisions.

IUCN Red List divides species into nine categories: Not Evaluated, Data Deficient, Least Concern, Near Threatened, Vulnerable, Endangered, Critically Endangered, Extinct in the Wild and Extinct. Critically Endangered, Endangered, and Vulnerable would be explained as follows:

(1) Critically endangered (CR). in a particularly and extremely critical state. Due to the habitat destruction, excessive exploitation or diseases and pests, these plants are difficult to restore to their original state even if measures are taken. They are few individuals and narrow distribution. For example, *Carpinus puoensis* is only one individual in nature in Putuoshan, Zhejiang province; there are seven individuals of *Osrya rehderiana* in Tianmu Mountains.

(2) Endangered (EN). very high risk of extinction in the wild, meets any of criteria A to E for Endangered. At present, Endangered plants are a certain number of individuals in a geographical area. However, because of artificial or natural reasons (deforestation, habitat degradation, excessive use, etc.), it can foresee its population is going to decrease in the near future. Typical examples are *Taxus* spp. and *Dalbergia odorifera*.

(3) Vulnerable (VU). meets one of the 5 red list criteria and thus considered to be at high risk of unnatural (human-caused) extinction without further human intervention.

0.1.5 CITES

CITES (Convention on International Trade in Endangered Species of Wild Fauna and Flora, or Washington Convention) is a multilateral treaty to protect endangered plants and animals. It was drafted as a result of a resolution adopted in 1963 at a meeting of members of IUCN. The convention was opened for signature in 1973 and CITES entered into force on July 1st, 1975. China jointed in the convention in 1981 formerly. Today, 182 countries and the European Union implement CITES, which accords varying degrees of protection to over 35000 species of animals and plants. Under this treaty, countries work together to regulate the international trade of animal and plant species and ensure that this trade is not detrimental to the survival of wild populations. Any trade in protected plant and animal species should be sustainable, based on sound biological understanding and principles. Each protected species or population is included in one of three lists, called appendices. The Appendix that lists a species or population reflects the extent of the threat to it and the controls that apply to the trade:

(1) Appendix I. species threatened with extinction. Trade in specimens of these species is

permitted only in exceptional circumstances. Notable plant species listed in Appendix I include *Ariocarpus* spp. (Cactaceae), *Agave parviflora* (Agavaceae), *Fitzroya cupressoides* (Cupressaceae), some species of Liliaceae and so on.

(2) Appendix II. species not necessarily threatened with extinction, but in which trade must be controlled in order to avoid utilization incompatible with their survival. Examples of species listed on Appendix II are *Panax ginseng* (Araliaceae), *Tillandsia harrissii* (Bromeliaceae), *Cyathea* spp. (Cyatheaceae), most parts of Cycadaceae, etc.

(3) Appendix III. species that are protected in at least one country, which has asked other CITES Parties for assistance in controlling the trade. Examples of species listed on Appendix III and the countries that listed them are the *Podocarpus neriifolius* by Nepal, *Bulnesia sarmientoi* by Argentina, *Dipteryx panamensis* by Costa Rica and Nicaragua, etc.

0.1.6 Old and notable trees

Old trees are keystone ecological structures that play important roles in supporting natural community structure and dynamics, maintaining critical ecosystem functions, and providing habitat for a wide range of native organisms. They have high conservation, cultural and aesthetic values. The old trees are classified into three types according to age: (i) grade one: over 500 years; (ii) grade 2: 300-499 years and (iii) grade 3: 100-299 years. Since the announcement of 'Technical Guidelines for Document Establishment of General Survey of National Ancient-Famous Trees' in 2001 in China, many cities have established official designations. The characteristics and distribution of heritage trees vary notably by geographical regions, management measures and urban development patterns. For example, Beijing has identified 39408 heritage trees, with *Platycladus orientalis*, *Pinus tabuliformis*, *Sabina chinensis* and *Sophora japonica* accounting for 94%. Most of them have been preserved in historical sites, temples and courtyards. In the Qiannan Prefecture of Guizhou Province, 5571 heritage trees were designated with only 238 in urbanized areas. *Liquidambar formosana* contributing 20% was the most dominant species. In Sanya, the southernmost city of China, the 968 heritage trees were dominated by *Tamarindus indica*, and most trees were associated with private rural residences. Therefore, there are rich in old trees resource all over the China, reflecting the long historic of our country.

Notable trees are not related to the tree age, but refer to the trees (i) planted by Chinese and foreign celebrities and leaders who have exerted great influence on history or society, and (ii) were extremely important history, cultural value or commemorative significance.

0.2 Forest resources and current status in China

0.2.1 Forest resources

China is in the front ranks of the world in terms of the total amount of forest resources. The forest area in China accounts for 5% of world's total, ranking the fifth behind Russia, Brazil, Canada and the United States. The forest stock volume is 3% of the world's total, ranking the sixth after Brazil, Russia, the United States, the Democratic Republic of the Congo and Canada. The plantation area ranks China the first in the world.

Based on the results of the National Forest Inventory, China has achieved the 'double increases' in forest area and forest stock volume since the early 1990s. In particular, after entering in the 21st century, with forest resources in a period of rapid growth, China has become one of the countries with the fastest growing forest resources in the world, and played a significant role in maintaining global ecological balance, conserving biodiversity, addressing climate change and promoting sustainable economic, ecological and social development.

However, China is still a country with a shortage of forest resources, fragile ecology and deficiency of eco-products, due to the impacts of nature, history, population and pressure of economic development. The forest coverage nationwide is far below the world average of 31%. The forest area per capita only makes up 1/4 of the world average, while the forest stock volume per capita accounts for merely 1/7 of the world average. The situation of insufficient forest resources with low quality and an uneven distribution has not been fundamentally changed, and the ecological security barrier has not been built in China.

China has only 5% of the world forest area and 3% of the world forest stock volume but would have to meet the huge demand for ecological and wood products from 23% of the world population, so that China has faced an increasing supply pressure. Strong protection and development of forest resources, promotion of sustainable forest management, increase of the total amount of forest resources, improvement of forest quality and ecological functions have become the strategic task of the Chinese government to promote ecological progress and build a beautiful country.

0.2.2 Forest vegetation types

China, with various types of landforms and complicated natural climate conditions, is rich in plant species and has diverse forest types with obvious zonal and distributional features. The main forest vegetation in China ranges from cold-temperate coniferous forests, temperate coniferous forests, temperate coniferous and broad-leaved mixed forests, warm coniferous forests and deciduous broad-leaved forests, to subtropical evergreen and deciduous broad-leaved mixed forests, evergreen broad-leaved forests, sclerophyllous evergreen broad-leaved forests, tropical monsoon forests and rainforests, which have made up a distinctive glorious forest landscape.

(1) Cold-temperate coniferous forests. They are distributed in the cold-temperate zone and the subalpine regions with the low and middle latitudes, covering a total area of 15.43 million ha. The forests are relatively concentrated in the hilly areas of northeast and north China, Qinba, Inner Mongolia and Xinjiang Autonomous Regions, the eastern and southern margins of Qinghai-Tibetan Plateau and Taiwan Province. Temperate coniferous forests are distributed in the plain areas, hilly land areas and low mountains in warm temperate zone, in the middle mountains in subtropical and

tropical zones, covering about 5 million ha. The forests are composed of three formation groups, namely temperate pine forests, Chinese arborvitae forests and Chinese cryptomeria forests. The temperate pine forests are distributed northward in the hilly area of north China, southwestward in Qinba hilly areas of Sichuan, Shaanxi and Hubei provinces, and eastward in the low and hilly land areas of the Huaihe River basin. The Chinese arborvitae forests are widespread in the areas of the northern part of China. The Chinese cryptomeria forests are mainly distributed in the mountainous areas of Zhejiang, Fujian and Jiangxi provinces.

(2) Temperate coniferous and broad-leaved mixed forests. They are the transitional forest vegetation from mountainous broad-leaved forests to mountainous coniferous forests, covering a total area of 5.04 million ha. The coniferous and broad-leaved mixed forests dominated by Korean pine are distributed in Changbai mountains, Laoye and Zhangguangcai mountains, Wanda mountains and the low and middle areas of the lesser Xing'an mountains in northeast China. The coniferous and broad-leaved mixed forests dominated by Chinese hemlock are distributed in the forest area of subalpine and middle mountains in southwest China.

(3) Warm coniferous forests. They are distributed in the subtropical low mountains, hilly and plain areas, covering a total area of 15.04 million ha. There are two types of forests, namely deciduous coniferous forests and evergreen coniferous forests. Among which the deciduous coniferous forests are mainly distributed in Sichuan, Hunan, and Hubei provinces and the evergreen coniferous forests grow northward in the areas of Qinling mountains, Funiu mountain and Huaihe River, southward in the areas of Baise city of Guangxi Autonomous Region, and Leizhou peninsula of Guangdong province, westward in the areas of Qingyi River basin in Sichuan province, and eastward in the areas of Taiwan province.

(4) Deciduous broad-leaved forests. They are distributed in northern temperate zone, covering an area of about 47.58 million ha. They consist of typical deciduous broad-leaved forests, mountainous poplar and birch forests and riparian deciduous broad-leaved forests. Among which typical deciduous broad-leaved forests are distributed in northeast and north China. Mountainous poplar and birch forests are widespread in the areas of northeast north, northwest and southwest China, while riparian deciduous forests can mostly be found along both sides of the rivers in the temperate zone with significant climate change.

(5) Evergreen and deciduous broad-leaved mixed forests. They are distributed in the northern part of subtropical region. The area of the mixed forests is about 7.74 million ha. There are three vegetation subtypes including deciduous and evergreen broad-leaved mixed forests, mountainous evergreen and deciduous broad-leaved mixed forests, and limestone evergreen and deciduous broad-leaved mixed forests. Which are distributed northward in the areas of Qinling mountains and Huaihe River, and stretch southward throughout the subtropical zone of the country.

(6) Evergreen broad-leaved forests. They are distributed in the subtropical region of China, covering about 13.45 million ha. They are mainly distributed in the areas of the south slope of Qinling mountains, Hengduan mountains, Yunnan-Guizhou Plateau, and low mountains, hilly and

plain areas of Sichuan, Hubei, Hunan, Guangdong, Guangxi, Fujian and Zhejiang provinces (Autonomous Region), the southern part of Anhui and Jiangsu Provinces as well as East China Sea islands and the northern part of Taiwan Province. Sclerophyllous evergreen broad-leaved forests are located in the southeast line of the Qinghai-Tibetan Plateau in the western and southwestern subtropical region and in the areas of Hengduan mountains, covering about 2.12 million ha. They are mainly distributed in the areas of Sichuan, Yunnan, Guizhou provinces and Xizang Autonomous Region.

(7) Monsoon forests. They grow in the tropical zone where the wet and dry seasons alternate periodically, covering about 0.22 million ha. They are distributed in the tropical region of six provinces (autonomous regions) of Taiwan, Guangdong, Guangxi, Yunnan, Hainan and Tibet.

(8) Rainforests. They are composed of tall and evergreen species in the high temperature and rainy tropical regions, covering about 0.59 million ha, and they can be found in Hainan Province, the southern part of Taiwan, Guangdong, Guangxi and Yunnan provinces (Autonomous Region) and the southeastern part of Tibet Autonomous Region.

0.2.3 Plant diversity in China

China is one of the richest countries in plant diversity, ranking third in the world (behind Brazil and Colombia) in number of species, and one of the world's 17 'mega-diversity' countries. The estimated number of vascular plant species may approach 33000, including 30000 angiosperms, 250 gymnosperms, and 2600 pteridophytes (up to 12%, 27% and 20% of world's total, respectively). Furthermore, approximately 2200 bryophytes can be found in China. There are more than 3000 genera and ca. 350 families of vascular plants. Nevertheless, these figures refer to mainland China and do not include either Taiwan or Hong Kong. Taiwan alone harbors more than 4000 vascular plants (over 3300 angiosperms, about 30 gymnosperms, and about 600 pteridophytes). With an area of only about 1100 km^2, Hong Kong still retains a very rich plant diversity, with more than 2100 higher plants.

China encompasses enormous diversity in geographical, climatological and topographical features, in addition to a complex and ancient geological history (with most of its lands formed as early as the end of the Mesozoic era). The country spans five major climatic zones (cold-temperate, temperate, warm-temperate, subtropical, and tropical), and is home to the highest mountain range on Earth (the Himalayas) and perhaps the most rugged one (the Hengduan mountains), vast plateaux such as the Tibetan (Qinghai-Xizang) Plateau, deserts (e.g. Taklamakan), deep depressions (e.g. Turpan Depression), large flat areas (e.g. Sichuan basin and North China plains), and some of Asia's largest rivers, including the Lancang River, Yarlung Zangbo River, Yangtze River, and Yellow River. All of these features contribute to the enormous diversity of ecosystems, including almost all types of forests, grasslands, shrublands, deserts (which cover more than 25% of the Chinese territory), marshes, savannas, tundras, or alpine meadows. Thus, it is not surprising that 19 of the 238 WWF global priority ecoregions are located totally or partially within China.

China possesses the richest flora of the North Temperate Zone. However, more relevant are the rates of endemism: 50% to 60% of the total number of species (i. e. 15000 to 18000) might be endemic to China. This wealth of species diversity and endemism is attributable to a series of factors largely related to the biogeography, tectonics and geological history of the country, including: (i) a complex and extended geological history, with many tectonic events, (ii) a large proportion of the area of China within tropical and subtropical latitudes, (iii) the wide and persistent connection of China to tropical regions of Southeast Asia as well as with other regions, (iv) an unbroken connectivity between tropical, subtropical, temperate, and boreal forests, (v) a highly rugged and dissected topography (especially in southern China), and (vi) perhaps the most significant, reduced extinction rates during the late Cenozoic global cooling.

0.2.4 Current status of endangered species in China

One of flora trait in China is its high level of endangerment. Most estimates show that 3000–5000 species could be threatened with extinction, i. e. up to 20% of the total flora. However, and according to more recent studies, this figure could be even higher: from the ca. 4200 angiosperm taxa (i. e. just 14% of the total number of angiosperms in China) assessed in the first phase of the elaboration of the China Species Red List, over 3600 (87%) were regarded as threatened following the 2001 IUCN criteria, and up to 651 were assigned to the highest endangerment category (CR, 'Critically Endangered'), that is, facing an imminent risk of extinction. These figures, nevertheless, should be taken with caution as the red list was biased toward rare and endangered species. At the beginning of the 2000s, it was estimated that, since the 1950s at least 200 plant species had become extinct; some conspicuous examples of these are *Otophora unilocularis* (not seen since 1935) and *Rhododendron kanehirai*, whose natural populations were flooded by a dam in Taiwan although it is extensively cultivated. However, some plant taxa have also been lost from the wild during the last decade (2000–2009), such as *Betula halophila*, *Cystoathyrium chinense* or *Plantago fengdouensis*. In addition to these losses, many species remain on the brink of extinction. An appreciable number of taxa are in an extreme situation of risk with population sizes often consisting of fewer than 100 individuals: examples include the gymnosperms *Cupressus chengiana* var. *jiangeensis* and *Abies beshanzuensis* var. *beshanzuensis* (from which only one and three individuals are remaining, respectively), and the angiosperms *Carpinus putoensis*, *Gleditsia japonica* var. *velutina*, and *Acer yangbiense* (with just one, two and four individuals remaining in the wild, respectively); these taxa are undoubtedly among the most endangered plants on the earth.

Endangered plant species tend to be concentrated in the southern part of the country, thus showing high congruence with both centres of species richness and centres of endemism. Thus, 'biodiversity hotspots' in its broadest sense (as centres of species richness, endemism and threatened species) are entirely located in the central and southern mountainous regions of China, which are moderately populated (much less compared to the North China Plains) and where the agricultural exploitation has been relatively limited due to their low suitability (too steep). However, other practices such as extensive logging and overgrazing have significantly damaged the natural ecosystems in these areas.

The main threats to Chinese plant diversity, both directly and indirectly, can be described as below: (i) habitat destruction, (ii) environmental contamination and global climate change, (iii) over-exploitation of species for human use, (iv) introduction of exotic species, (v) lack of effective environmental scope of government policies and ineffective legal protection, (vi) economic and population growth.

0.3 Legislation conservation of plant diversity in China

Environmental legislation and government planning (i.e. policies) are also essential to ensure adequate conservation of biodiversity. China has issued various laws and regulations addressing biodiversity conservation since the early 1980s. The most relevant laws governing plant biodiversity include the *Environmental Protection Law* (issued in 1979, revised in 1989), the *Forest Law* (issued 1984, revised 2020), the *Grassland Law* (issued 1985, revised 2002), and the *Seeds Law* (2000, revised 2004). China has also issued a significant number of regulations and rules, such as the *Regulation about Protection and Administration of Wild Medicinal Material Resources* (1987), the *Regulation about Nature Reserves* (1994), the *Regulations on Wild Plants Protection* (1996), or the *Regulation on the Import and Export of Endangered Wild Fauna and Flora Species* (2006) and *Regulation on Scenic Spots and Historical Sites* (2006). In addition to laws and regulations, there is some governmental supervision actively supporting biodiversity conservation in China, the most relevant being the *Environmental Impact Assessment* (EIA) system, which was legally implemented in 1981 and amended several times thereafter (1986, 1989, 1998, and 2002). Other mechanisms include licensing systems (such as forest logging and land use licenses), economic incentives (financial subsidies, tax-deductions, and compensation fees to enhance sustainable exploitation of natural resources, and more recently, payment for ecological and environmental services), or the quarantine system (established in the early 1980s).

Regarding government planning, China started to launch several comprehensive biodiversity-related policies from the early 1990s. Within the *Eighth Five-Year Plan for Economic and Social Development* (1991-1995), China took biodiversity conservation as a key national policy. In addition to starting an inventory of biodiversity at all levels, the *China Biodiversity Conservation Action Plan* was launched in 1994 to implement the *Convention on Biological Diversity* (CBD) together with *China's Agenda* 21. The *Ninth Five-Year Plan* (1996-2000) formally included the execution of CBD: *China's Biodiversity: a Country Study* plan was launched at the end of 1997, which analyzed the country's overall biodiversity, its economic value and benefits, the cost of implementing the CBD, and long term objectives for biodiversity conservation and sustainable use of biological resources. Other major plans issued before the end of the 20th century encompassed the *National Program for Nature Reserves* (1996-2010) and the *National Plan for Ecological Construction* (1998-2050).

At the turn of the century, new environmental policies were launched to cope with the need for a more comprehensive and sustainable nature management. These new policies, commonly known as the 'Six Key Forestry Programs' (SKFP), meant an investment which exceeded the total expenditure during the period 1949-1999, and were aimed to avoid some of the pitfalls of the past in nature management. The SKFP, launched essentially for ecosystem rehabilitation, environmental protection and afforestation, covered more than 97% of China's counties, and included: (i) the National Forestry Protection Program (NFPP); (ii) the Shelterbelt Development Program (SDP); (iii) the Grain to Green Program (also known as the Sloping Land Conversion Program and the Cropland to Forest Program) (GTGP); (iv) the Sand Control Program for areas in the vicinity of Beijing and Tianjin (SCP), (v) the Wildlife Conservation and Nature Reserves Development Program (WCNRDP); and (vi) Fast-growing and High-yielding Timber Plantations Program (FHTPP). Most of these programs were planned to expire in 2010 except the last one, which will be alive until 2015. At the end of 2010 the *China National Biodiversity Strategy and Action Plan* (2011-2030) was approved to replace the 1994 plan. Specific to plant biodiversity, in 2007 the *China's Strategy for Plant Conservation* (CSPC) within the *Global Strategy for Plant Conservation* (GSPC) was launched, aiming to pursue the CBD 2010 Biodiversity Target.

All these plans have also been designed to fit other international treaties and conventions with implications for plant diversity signed by China, in addition to the CBD: the *CITES Convention* (1981), the *Ramsar Convention* (1992), the *United Nations Convention to Combat Desertification* (UNCCD) (1996), the *UN Millennium Development Goals* (2000), the *Kyoto Protocol* (2002), and the *Cartagena Protocol on Biosafety* (2005), among others. In addition, China also maintains international cooperation with governments and both public and private institutions, highlighting: the 'China Council for International Cooperation on Environment and Development', the 'China-EU Biodiversity Project', the 'Greater Mekong Subregion Core Environment Program, and the 'Sino-American joint investigation on plant diversity in Hengduan mountains'.

0.4 Botany subdisciplines

Botany, plant science(s), phytology, or plant biology is a branch of biology and is the scientific study of plant life and development. Botany covers a wide range of scientific disciplines that study plants, algae, and fungi including: structure, growth, reproduction, metabolism, development, diseases, and chemical properties and evolutionary relationships between the different groups. Botany, the study of plants, began with tribal efforts to identify edible, medicinal and poisonous plants, making botany one of the oldest sciences.

As with other forms of life, plants can be studied at many different levels. One is the molecular level, which is concerned with the biochemical, molecular and genetic functions of plants. Another is the cellular, tissue and organelle (a discrete structure of a cell that has a specialized function) level, which studies the anatomy and physiology of plants; and the community and population level, which involves interactions within a species, with other species and with the environment. Thus, botany can be considered to comprise eight key sub-disciplines each studying a different aspect of plants. These disciplines are genetics, systematics, cytology, anatomy, morphology, physiology, pathology, and ecology.

(1) Genetics is the study of heredity, genes, and gene function. Much modern botany has put to use plant DNA and genomic information to study plants more rigorously than it was previously possible. Molecular biology has allowed taxonomists to categorize plant species based on DNA. Plants have been classified into different families and renamed as a result. For this reason, older botanical guides may contain outdated names and classifications. A considerable amount of new knowledge today is being generated from studying model plants like *Arabidopsis thaliana* (mustard weed).

(2) Botanical systematics is the study of plant characteristics, especially for the purpose of discerning their evolutionary relationships and establishing different plants' phylogenetic associations. The term 'systematics' may or may not overlap with 'taxonomy', which concerns itself with scientific classification of species and other taxa. Recent developments are cladistics and molecular systematics. Actually, this course is one of branches for plant taxonomy.

(3) Cytology is the study of cells, including their function, structure, and life history.

(4) Anatomy is the study of the interior structure of living things.

(5) Morphology is the study of the exterior form of plants, including the placement of stems and leaves on a stem (i.e. alternate or opposite), and also the study of life histories and evolutionary development. Botanical field guides often rely on plant morphology to help biologists identify plant species in the field. In the course, we will identify the various plant species by different morphological characteristics.

(6) Plant physiology is the study of the function of plants and their cells and tissues. Examples of physiological research includes the study of photosynthetic pathways in different plants and mineral uptake by plants.

(7) Plant pathology is the study of diseases and the structural and functional changes that occur with diseases. This can be important for range of fields, including conservation biology, ecology, agriculture, and horticulture.

(8) Plant ecology is the study of interactions between organisms and their biotic and abiotic environment as an integrated system.

Other sub-disciplines of botany include: (i) Ethnobotany, the study of how a particular culture, or region has made use of local and indigenous plants, including their use for food, shelter, medicine, clothing, hunting and religion. (ii) Paleobotany is the study of fossil plants. Palynology, the study of modern and fossil pollen, is also often grouped with paleobotany. Paleobotany and palynology are both instrumental in studying paleoclimatology. (iii) Bryology is the study of mosses, liverworts, and hornworts; phycology is the study of algae; pteridolgy is the study of ferns; and mycology is the study of fungi.

In reviewing the sweep of botanical subdisciplines, it is evident that, through the power of the scientific method, most of the basic questions concerning the structure and function of plants have, in principle, been resolved. Now the distinction between pure and applied botany becomes blurred as our historically accumulated botanical wisdom at all levels of plant organization is needed (but especially at the molecular and global levels) to improve human custodianship of planet earth. The most urgent unanswered botanical questions now relate to the role of plants as primary producers in the global cycling of life's basic ingredients: energy, carbon, hydrogen, oxygen, and nitrogen, and ways that our plant stewardship can help address the global environmental issues of resource management, conservation, human food security, biologically invasive organisms, carbon sequestration, climate change, and sustainability.

0.5 Aims for the course

(1) To know how to collect specimens. When collecting plants, get all plant parts in as many stages of development as possible. Buds, leaves, flower parts and fruits will aid in identification. If a plant has a unique characteristic that distinguishes it from other species of the same genus, be certain to include this feature with the collected specimen.

(2) To know how to prepare specimens for future preservation. After the plant specimens have been collected, and handled. They place specimens in the presses while they are out in the field. This practice prevents wilting and insures preservation of the plant's shape and color. As it may be difficult to carry ventilators to the field when collecting, specimens can be pressed between newspapers in the field.

(3) To know how a manual should be used. It is better to select the local flora, such as *Jiangsu Flora*, *Huangshan Flora*, etc., the more specific flora, the better to identify the aimed plant.

(4) To know how to use identification keys. The key consists of a series of choices, based on observed features of the plant specimen. It provides a choice between two contradictory statements resulting in the acceptance of one and the rejection of the other. A single pair of contradictory statements is called a couplet and each statement of a couplet is termed a lead. By making the correct choice at each level of the key, one can eventually arrive at the name of the unknown plant.

(5) To recognize divisions, classes, orders, families, genera, and species. It is the basic content of plant taxonomy for students to acknowledges, and they need to know the biological hierarchy.

(6) To know how the plants are described. The first step on identifying plant species is understanding the language of plants —the language used to describe the many forms that plants and plant parts come in morphology, especially: root, stem, leaf, flower, and fruit.

(7) To become familiar with the basic taxonomic principles, and with at least one system of plant classification. In the textbook, Zheng's Classification System will be used for gymnosperm, and Hutchinson Classification System for angiosperm.

References:

New World Encyclopedia contributors. Botany, New World Encyclopedia[EB/OL]. [2020-04-23]. https://www. newworldencyclopedia. org/p/index. php? title=Botany&oldid=1005911.

Terri Schab. What Are Five Different Fields of Botany? [EB/OL]. (2017-04-24) [2020-04-23]. https://sciencing. com/five-different-fields-botany-16728. html.

Biocyclopedia. Botany Subdisciplines[EB/OL]. [2020-04-23]. https://biocyclopedia. com/index/botany_subdisciplines. php.

North Carolina Association for Biomedical Research. Botany [EB/OL]. [2020-04-23]. https://www. aboutbioscience. org/topics/botany/

BYJU'S. What is Plant Taxonomy [EB/OL]. [2020-04-23]. https://byjus. com/neet/important-notes-of-biology-for-neet-plant-taxonomy/

Sharma O P, 1993. Plant taxonomy[M]. New-Delhi: Tata McGraw-Hill Education.

Sivarajan V V, 1991. Introduction to the principles of plant taxonomy [M]. Cambridge: Cambridge UniversityPress.

Queensland Herbarium, 2016. Collection and preserving plant specimens, a manual[M]. 2nd edition. Brisbane: Department of Science, Information Technology and Innovation.

López-Pujol J, Wang H F, Zhang Z Y, 2011. Conservation of Chinese plant diversity: an overview[J]. Research in biodiversity-Models and applications. InTech, Rijeka, 163-202.

Caldwell M M, 1981. Plant response to solar ultraviolet radiation. Physiological plant ecology I. Encyclopedia of Plant Physiology New Series (Vol. 12A) [M]. Berlin: Springer-Verlag.

Chapin S F, Matson P A, Mooney H A, 2002. Principles of Terrestrial Ecosystem Ecology[M]. New York: Springer.

Chapter 1

Morphological Features of Forest Plants

China is one of the richest countries for plant diversity with approximately 33000 vascular plant species, ranking second in the world. Many of them play an important role in the field of medicine, gardening, ecological conservation, commercial timber, and so on. However, the plant diversity in China is increasingly threatened, not only the extinction or environment destruction, but also crime. Plant morphology is a scientific discipline in the visual identification of plants, which can help us to identify a plant using by flower, fruit, leaf or any other morphology quickly and accurately. The following morphological terms in this chapter are used repeatedly in the description and identification of plants.

1.1 Root

It is a part of the plant axis that mostly grows down into the earth and is concerned with absorption of water and minerals. There are important internal structural differences between stems and roots. Because of roots erupting in different part, they can be divided into three types: primary root (主根), secondary roots (侧根), and adventitious roots (不定根) (Figure 1-1).

To adapt to the environment, many plants have to become specialized to serve adaptive purposes, for example aerating roots, aerial roots, storage roots,

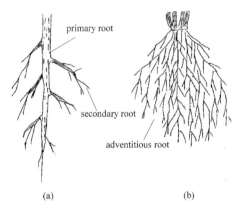

Figure 1-1 Types of root
(a) tap root system; (b) adventitious root system.

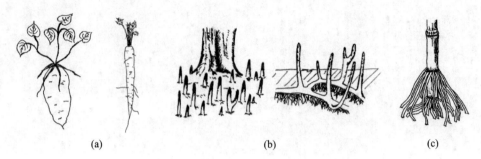

Figure 1-2　Main types of modification root
(a) storage tap root; (b) pneumatophore; (c) prop root.

propagative roots, and so on. Some of modification roots are shown in Figure 1-2.

1.2　Stem

A stem is the main structural axes in vascular plant. It supports leaves, flowers and fruits, transports fluids between the roots and the shoots in the xylem(木质部) and phloem(韧皮部), stores nutrients, and produces new living tissue. For the tall arbor species, commercial timbers also come from this part.

1.2.1　Habit

(1) Woody plant(木本植物)

Tree (arbor)(乔木): a woody perennial with a main stem, for example: pine, aspen, birch, etc.

Shrub(灌木): a low woody plant with several stems arising from or near the ground, such as jasmine.

Woody liana(木质藤本): a long-stemmed, more flexible growth at the base of the stem, for instance: Chinese wistaria.

(2) Herb(草本植物): a small plant with soft stem-annual, biennial or perennial, without a woody stem.

1.2.2　Twig

Twig (枝条) is a small, thin branch of a tree or bush, which provides a brilliant meanings of identification, especially for the deciduous species in winter. The most conspicuous features of twigs are their buds, leaf scars, stipulate scars, lenticels, and pith; in addition, their color, taste, odor, the presence or absence of hairs, spur shoots, and thorns are also valuable characteristics for aims of identification. The general twig feature is shown in Figure 1-3.

(1) Bud(芽). It is a small but conspicuous part on twig, and it develops into a flower or leaf in the spring mostly. According to their location, status, morphology, and function, they can be categorize into: (i) terminal buds, lateral (axillary) buds, and adventitious buds; (ii) accessory

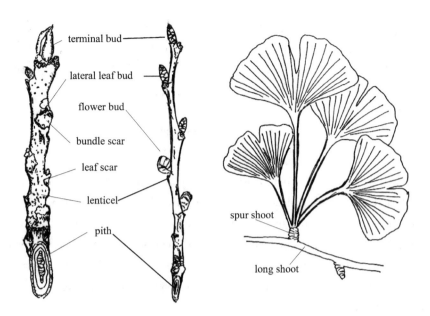

Figure 1-3 Twig feature for the common woody plant

buds, resting buds, dormant or latent buds, and pseudoterminal buds; (iii) scaly buds and naked buds; (iv) vegetative buds, reproductive buds, and mixed buds (Figure 1-4).

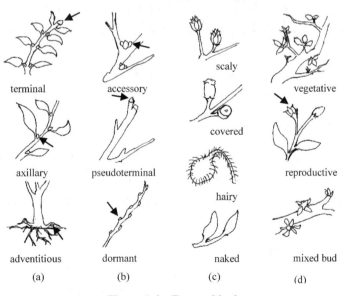

Figure 1-4 Types of bud

(2) Pith(髓). It is a tissue (composed of soft, spongy parenchyma cells) in the stems of vascular plants, which is the medial or central portion of a twig. The form and composition of pith are very useful to determine some species in the temperature zone. Usually, we can find four kinds of piths in nature (Figure 1-5).

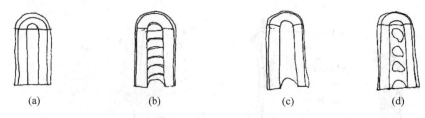

Figure 1-5 Types of pith
(a) uniform; (b) chambered; (c) hollow; (d) excavated.

(3) Bark(树皮). It is the outermost part of branch or trunk, which includes inner bark (living tissue) and outer bark (dead tissue); actually, the bark we talking about as usually indicate the outer one. Bark is also one of the most important features in the identification of large trees, especially for logs when cut down and without any leaves and twigs. The most common barks seen in daily are showed in Figure 1-6. But we have to be aware that the features of bark are various from seedling to adult tree sometimes, such as color or surface. Therefore, it is an empirical characteristic not absolute.

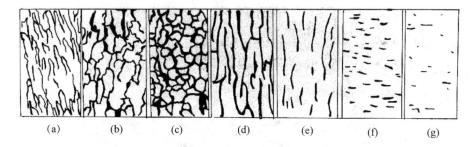

Figure 1-6 Types of bark
(a) strips; (b) scales; (c) cracks; (d) ridges; (e) furrows; (f) lenticels; (g) smooth.

1.2.3 Branching type

Due to the various activity of terminal buds and lateral buds in the main stem, different species present different branching, which include 3 types (Figure 1-7): monopodial branching (单轴分枝)[Figure 1-7(a)], sympodial branching(合轴分枝)[Figure 1-7(b)], dichotomous branching (假二叉分枝)[Figure 1-7(c)].

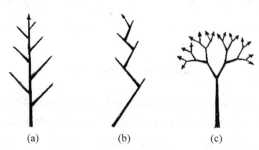

Figure 1-7 Branching types
(a) monopodial branching; (b) sympodial branching; (c) dichotomous branching.

1.2.4 Modification of stem

Stem modifications, either aboveground, underground, or aerial, enable plants to survive in particular habitats and environments. Some of the most important types of modifications of stem are as follows: (i) aerial modifications of stem (Figure1-8), (ii) underground modifications of stem (Figure1-9), and (iii) subaerial modifications of stem.

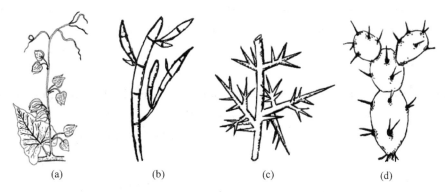

Figure 1-8 Types of aerial modifications of stem
(a) tendril; (b) cladodes; (c) horns; (d) phylloclade.

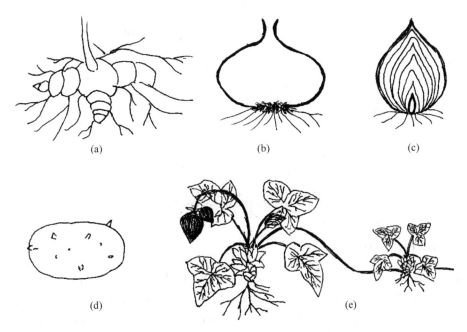

Figure 1-9 Types of subaerial modifications stem
(a) rhizome; (b) corm; (c) bulb; (d) tuber; (e) runner.

1.3 Leaf

A leaf is lateral photosynthetic organ of shoot. Its functions are photosynthesis, respiration, transpiration, and synthesis of secondary chemicals. Generally, leaves stay much longer than flowers and fruits in the plant, and it is important for identification. The feature of leaves, including (i) structure, (ii) form or general outline, (iii) venation, (iv) apex, (v) base, (vi) margin, (vii) position and arrangement, (viii) surface, and (ix) texture, are valuable to describe the species.

1.3.1 Structure

A typical leaf is composed of three parts (Figure 1-10): leaf blade, petiole, and stipule. If a leaf with three parts mentioned above is complete leaf, or is incomplete leaf. For example, the leaves of pear, peach and rose are complete; but just like the lilac tree and cabbage are without stipule—incomplete leaf.

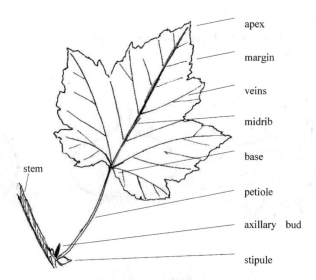

Figure 1-10 A structure of complete leaf

1.3.2 Morphology

While common leaf characters do not necessarily imply kinship among plants, leaf traits are nevertheless valuable in plant identification. External leaf characteristics, such as shape, margin, hairs, the petiole, and the presence of stipules and glands, are frequently important for identifying plants to family, genus or species levels. The most important thing is that they are easier to collect.

(1) leaf shape(叶形). For the same species, the leaf shapes keep in a certain extent. Some species have a special leaf shapes, such as ginkgo, tulip tree, palm. Hence, we can identify the

plant based on the stale leaf shape. The common leaf shapes include linear, lanceolate, oblanceolate, rotund (orbicular), elliptical (oval), ovate, obovate, oblique, oblong, reniform, cordate, and so on (Figure 1-11).

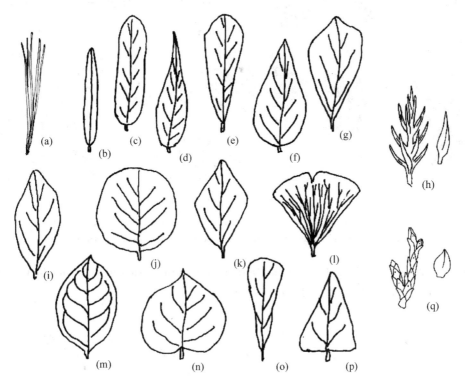

Figure 1-11　Basic leaf shapes

(a) needle-like (acicular); (b) linear; (c) oblong; (d) lanceolate; (e) oblanceolate;
(f) ovate; (g) obovate; (h) subulate; (i) elliptical; (j) orbicular; (k) rhombic;
(l) flabellate; (m) oval; (n) reniform; (o) spatulate; (p) deltate; (q) scale-like.

(2) leaf apex (tip) (叶尖). It is top of the leaf, and common types are obtuse, acute, acuminate, caudate, cuspidate, truncate, aristate, and emarginate (Figure 1-12).

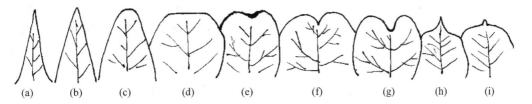

Figure 1-12　Basic leaf apices

(a) acuminate; (b) acute; (c) obtuse; (d) truncate; (e) retuse; (f) emarginate;
(g) obcordate; (h) cuspidate; (i) mucronate.

(3) leaf base (叶基). It is base of the leaf, and the base of the leaf blade could be attenuate, rounded, truncate (straight), cuneate, cordate and so on (Figure 1-13).

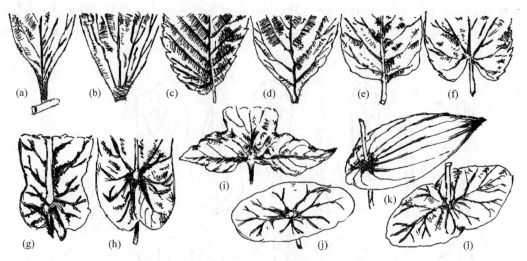

Figure 1-13 Basic leaf bases
(a) attenuate; (b) cuneate; (c) oblique; (d) obtuse; (e) truncate; (f) cordate; (g) auriculate;
(h) sagittate; (i) hastate; (j) peltate; (k) perfoliate; (l) connate-perfoliate.

(4) leaf margin(叶缘). Leaf margin variants are entire (smooth) and toothed: dentate, serrate, double serrate and crenate (Figure 1-14).

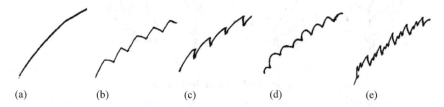

Figure 1-14 Basic leaf margins
(a) entire; (b) dentate; (c) serrate; (d) crenate; (e) double-serrate.

(5) leaf lobe(叶裂). The leaf margin is divided into many lobes, deeply than common serrate. There are three kinds of leaf lobes: trifid, palmately and pinnately (Figure 1-15).

(6) leaf venation(叶脉). The arrangement of vascular bundles or veins in a leaf. One is dichotomous venation (二叉脉), such as ginkgo; second is reticulate venation (网状脉), including pinnately (羽状脉), palmately (掌状脉), and ternately (三出脉); third is parallel venation (平行脉), including unicostate venation (侧出平行脉), divergent venation (射出平行脉), arcuate venation (弧形脉) and convergent venation (直出平行脉) (Figure 1-16).

(7) leaf texture(叶质地). It is presented by overall structure. There are three kinds of leaf texture: coriaceous (革质), membranous (膜质或纸质), and fleshy (succulent)(肉质).

(8) leaf arrangement(叶序). Position and arrangement of leaves are sources of fundamental features, often more constant than most others. There are including alternate (互生), opposite (对生), whorled (轮生), fasciculate (束生) and cluster (簇生) (Figure 1-17).

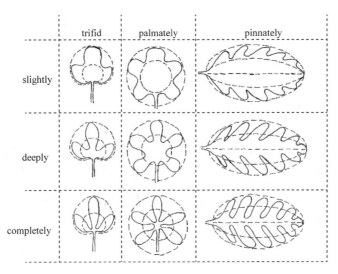

Figure 1-15 Basic leaf lobes

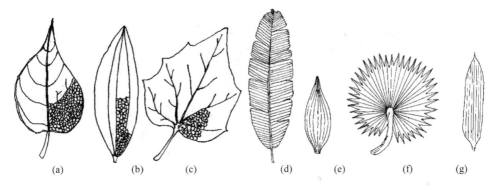

Figure 1-16 Types of venation

Reticulate venation: (a) pinnately; (b) palmately; (c) ternately.
Parallel venation: (d) unicostate; (e) arcuate; (f) divergent; (g) convergent.

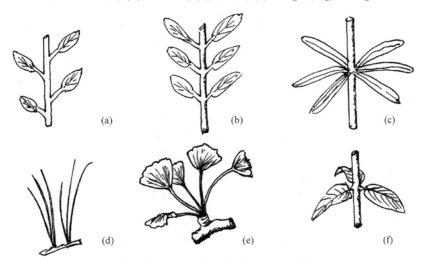

Figure 1-17 Types of leaf arrangement
(a) alternate; (b) opposite; (c)(f) whorled; (d) fasciculate; (e) cluster.

(9) Simple and compound leaf(单叶与复叶). A leaf with one blade is generally a simple leaf (单叶)(Figure 1-10); a leaf of two or more apparent blades is a compound leaf (复叶) and each apparent blade is a leaflet (小叶). There is not lateral bud in the axil of leaflet. There are four kinds of compound leaf: pinnately(羽状复叶), palmately(掌状复叶), unifoliate(单身复叶), and trifoliate(三出复叶) (Figure 1-18).

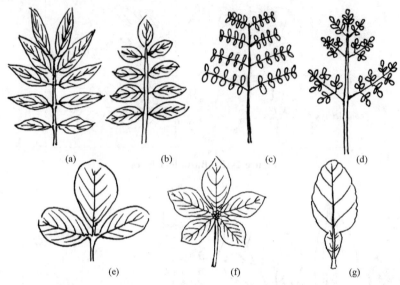

Figure 1-18　Types of compound leaf

Pinnately: (a) paripinnate; (b) imparipinnate; (c) bipinnate; (d) tripinnate.
Pamately: (e) trifoliate; (f) multifoliate; (g) unifoliate.

1.3.3　Modification of leaf

Leaves are specialised to perform photosynthesis. In addition, they also have other significant roles to play, such as support, storage of food, defence, etc. For each of these functions, they have been modified into different forms. For example, tendrils of peas, spines of cacti, leaves of insectivorous plants, etc. are different modified leaves (Figure 1-19).

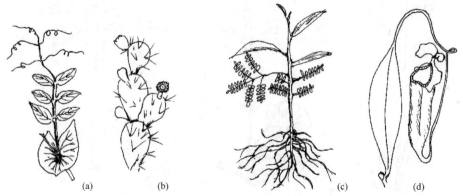

Figure 1-19　Types of modification leaf

(a) leaflet tendrils; (b) leaf spine; (c) phyllode; (d) insectivorous leaf.

1.4 Flower

Flowers are the reproductive part of a plant, but only angiosperms have the real flowers, called flowering plants or seed plants. As the flowers develop, the ovaries may grow into fruits. Flowers attract pollinators, and fruits encourage animals to disperse the seeds. They are not only involved in reproduction but are also a source of food for other living organisms. Actually, flowers are used extensively by botanists to identify and establish relationships among plant species due to their stability structure.

1.4.1 Structure

A flower is a compact generative shoot that is comprised of five parts: pedicel (花柄), receptacle (花托), sterile (perianth) (花被), male (androecium) (雄蕊), and female (gynoecium) (雌蕊). Perianth is typically split into green part (calyx, consists of sepals) and color part (corolla, consists of petals). Flowers can either be complete or incomplete. A complete flower is the one that consists of sepals, petals, stamens and pistil. On the contrary, an incomplete flower is the one that lacks one or more structures. A complete flower consists of two different parts: vegetative part and reproductive part. A typical flower is shown in Figure 1-20.

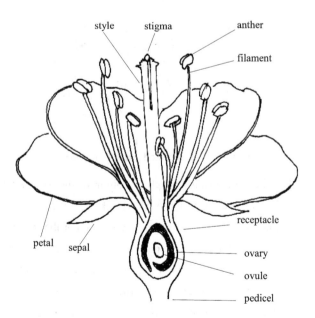

Figure 1-20 A typical flower structure

1.4.2 Types of flower

The general features that a flower has are sex, merosity, perianth, corolla, symmetry, the position of the gynoecium, and so on. According to the features we have mentioned above, the types of flowers can be divided into: (i) monochlamydeous, dichlamydeous and achlamydeous; (ii) bisexual, unisexual and asexual; (iii) polypetalous and gamopetalous.

1.4.3 Morphology of main parts

(1) Receptacle(花托). It is a vegetative tissue near the end of reproductive stems that are situated below or encase the reproductive organs. Generally, small individual flowers are arranged on a round or dome-like structure that is also called receptacle.

(2) Calyx (sepal)(花萼). It is the outermost whorl of a flower. It is consist of sepals (萼片), presenting at the base of a flower. These protect the flower whorls against mechanical injuries and desiccation. The development and form of the sepals vary considerably among flowering plants. After flowering, most plants have no more use for the calyx which withers or becomes vestigial. Some plants retain a thorny calyx, either dried or live, as protection for the fruit or seeds. In Figure 1-21, there is calycles (副萼) in *Hibicus*.

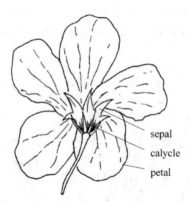

Figure 1-21   Calyx and calycle in *Hibicus*

(3) Corolla(花冠). They are usually brightly coloured and scented to attract pollinators. Corolla is consist of several petals in one to more whorls. Petals are modified leaves that surround the reproductive parts of flowers. There are various types of corolla in the form and color, but usually it is basically stable in different taxonomic level. Hence, it is very important to utilize corolla to identify the genera or family, such as cruciform, rosaceous, urceolate, campanulate (bell-shaped), papilionaceous (butterfly-like), bilabiate (two-lipped), and so on (Figure 1-22).

(4) Androecium(雄蕊群). It a male reproductive structure, consisting of many stamens (雄蕊). Stamen is also known as the third whorl of the flower and it consists of a filament (花丝) which is a thread-like structure with a circular structure anther (花药) on the top. Pollen is produced by the anther which contributes to the male reproductive process of the plant. When the

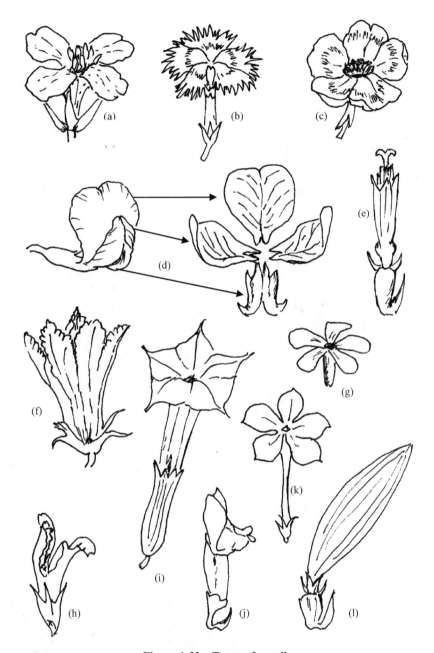

Figure 1-22　Types of corolla

(a) cruciform; (b) caryophyllaceous; (c) rosaceous; (d) papilionaceous; (e) tubular; (f) campanulate; (g) rotate; (h) bilabiate; (i) infundibuliform; (j) personate; (k) hypocrateriform; (l) ligulate.

stamens are united by their filaments only (anthers remain free), it is called adelphous condition (离生雄蕊), and when the stamens are united by their anthers only (filaments remain free), it is called syngenesious condition (聚药雄蕊). If, however, the filaments as well as the anthers of different stamens become united, it is called synandrous condition (单体雄蕊) (Figure 1-23).

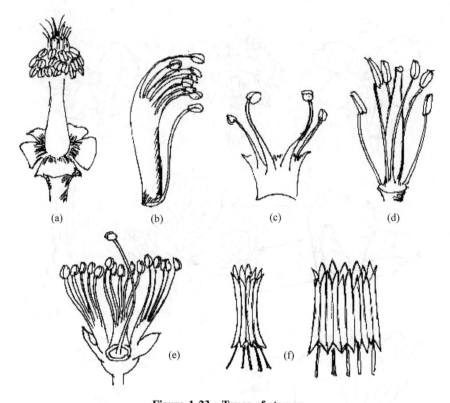

Figure 1-23　Types of stamen
(a) monadelphous; (b) diadelphous; (c) didynamous; (d) tetradynamous;
(e) polydelphous; (f) synantherous.

(5) Gynoecium(雌蕊). The fourth or female whorl composed of one or more carpels (心皮). The carpel locates in the center of flower. The carpels contain the pistil (雌蕊), the female reproductive part of the flower. It comprises of the ovary (子房), style (花柱), and stigma (柱头). Position of ovary (Figure 1-24), cohesion of carpels (Figure 1-25), and placentation (Figure 1-26) should provide useful information to identify the plants.

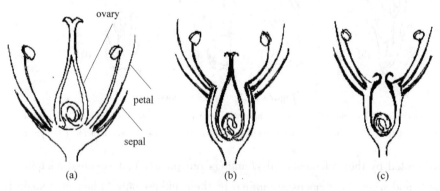

Figure 1-24　Position of ovary on receptacle
(a) hypogynous; (b) perigynous; (c) epigynous.

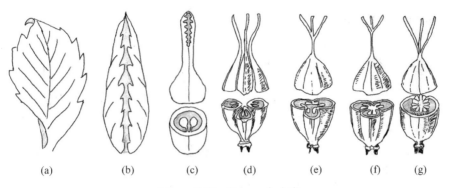

Figure 1-25 Types of pistil
(a)(b): evolution of carpel; (c) simple; (d)~(g): syncarpous.

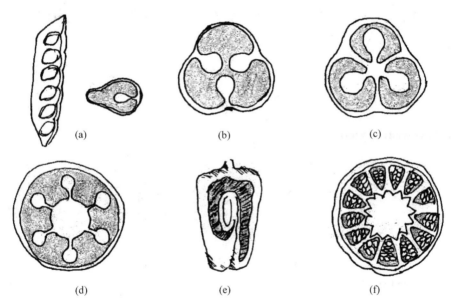

Figure 1-26 Various types of placentation
(a) marginal; (b) parietal; (c) axile; (d) free central; (e) basal; (f) superficial (or laminar).

1.4.4 Inflorescence

Generally, one flower locates in the axil or terminal of twigs, such as peony, rose, etc., called solitary. But there are many flowers on a special floral axis usually, called inflorescence. An inflorescence is a group or cluster of flowers arising from a branched or unbranched axis with a definite pattern. Inflorescences are described by many different characteristics including how the flowers are arranged on the peduncle, the blooming order of the flowers and how different clusters of flowers are grouped within it. These terms are general representations as plants in nature can have a combination of types. There are three main types of inflorescence-racemose (无限花序) (Figure 1-27 A), cymose (有限花序) (Figure 1-27 B), and special type (特殊类型) (Figure 1-27 C).

A: Inflorescence-racemose

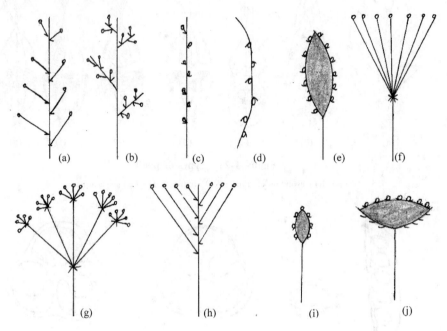

B: Inflorescence-cymose

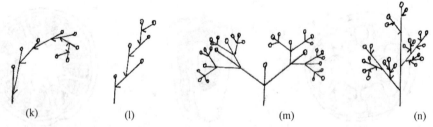

C: Inflorescence-special

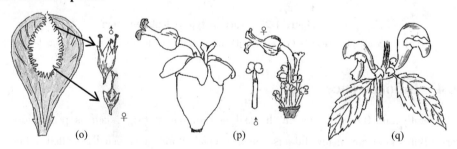

Figure 1-27　Types of inflorescence

A: (a) raceme; (b) compound raceme; (c) spike; (d) catkin; (e) spadix; (f) umbel; (g) compound umbel; (h) corymb; (i) head; (j) capitulum;

B: (k)~(l): monochasial cymes; (m) dichasial cyme; (n) panicle;

C: (o) hypanthodium; (p) cyathium; (q) verticillaster.

1.5 Fruit

Fruit is a mature or ripened ovary from flower. Fruits and seeds develop from flowers after completion of two processes namely pollination and fertilization. After fertilization, the ovary develops into fruit. The fruits of various species of plants exhibit a variety of forms, and fruit form is a valuable trait in classifying plants. According to the each part of flower development into fruit, they are classified into three types: true fruit (真果), false fruit (假果) and parthenocarpic fruit (无籽果).

1.5.1 Structure

A fruit consists of two main parts—the seeds and the pericarp or fruit wall. The structure and thickness of pericarp varies from fruit to fruit. The origins of the fruit coat and the pericarp which consists of three layers—outer exocarp (外果皮), middle mesocarp (中果皮) and inner endocarp (内果皮), are mostly from the wall of the pistil (Figure 1-28).

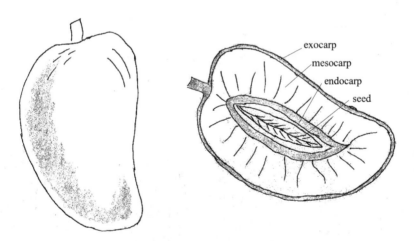

Figure 1-28 Fruit structure

1.5.2 Types

(1) A single fruit (单果). It is developing from a single ovary of a single flower with or without accessory parts. The ovary may be monocarpellary or multicarpellary syncarpous. On the nature of pericarp, simple fruits are divisible into two types: fleshy fruits (Figure 1-29) and dry fruits. Generally, dry fruits can be grouped into two assemblages: those that split open at maturity (dehiscent) (Figure 1-30) and those that do not (indehiscent) (Figure 1-31).

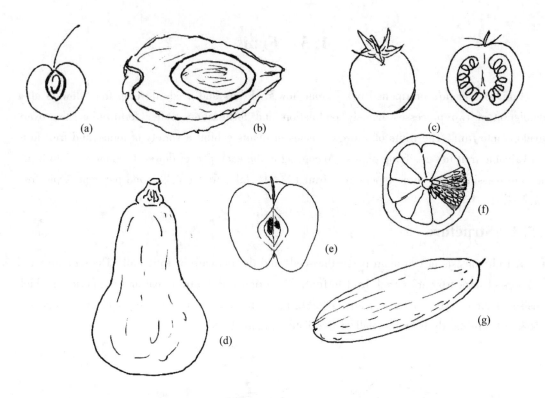

Figure 1-29　Types of fleshy fruits
(a) (b) drupe; (c) berry; (d) (g) pepo; (e) pome; (f) hesperidium.

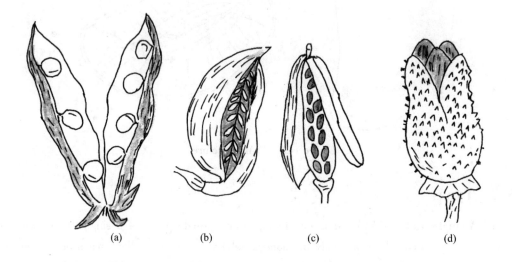

Figure 1-30　Types of dehiscent dry fruits
(a) legume (pod); (b) follicle; (c) silique; (d) capsule.

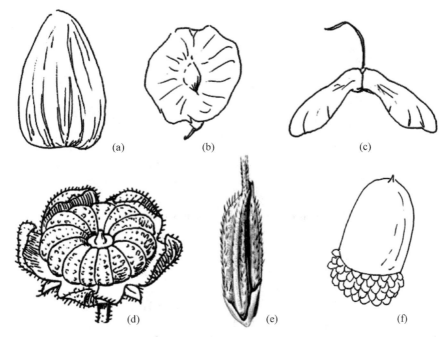

Figure 1-31　Types of indehiscent dry fruits
(a) achene; (b) (c) samara; (d) schizocarp; (e) caryopsis; (f) nut.

(2) Aggregate fruit (聚合果). Fruit develops from a single flower having several apocarpous pistils, all of which ripen together; finally, each of the free carpel is develops into a simple fruitlet. An individual ovary develops into a drupe, achene, follicle or berry. An aggregate of these fruits borne by a single flower is known as an etaerio (Figure 1-32).

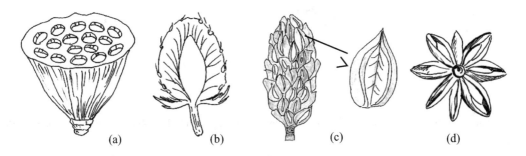

Figure 1-32　Types of aggregate fruits
(a) aggregate nut; (b) aggregate achene; (c) (d) aggregate follicles.

(3) Multiple of composite fruit (聚花果). Fruit is composed of several closely associated fruits, derived from entire inflorescence and forming one body at maturity (Figure 1-33).

There is an outline diagram for the fruit types as following, which would help you to classify the fruit quickly and clearly.

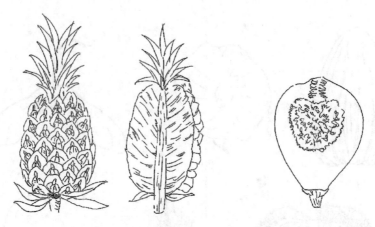

Figure 1-33 Types of multiple fruits

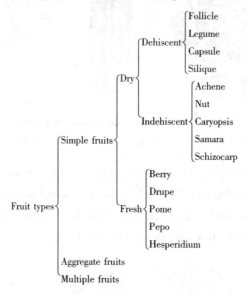

References:

Gurcharan Singh, 2019. Plant systematics: an integrated approach [M]. 4th edition. Boca Raton: CRC Press.

Mauseth James D, 2016. Botany: an introduction to plant biology [M]. 6th edition. Burlington, Massachusetts: Jones & Bartlett learning.

Dirk Walters, David J Keil, Zack E Murrell, et al., 2005. Vascular plant taxonomy [M]. 5th edition. Dubuque: Kendall HuntPublishing Company.

哈里斯, 王宇飞, 2001. 图解植物学词典[M]. 北京: 科学出版社.

方炎明, 2006. 植物学[M]. 北京: 中国林业出版社.

汪劲武, 2009. 种子植物分类学[M]. 北京: 高等教育出版社.

金银根, 2010. 植物学[M]. 2版. 北京: 科学出版社.

杜凤国, 王柏青, 2008. 森林植物学[M]. 长春: 吉林大学出版社.

Chapter 2

Foundamentals of Plant Taxonomy

Plant taxonomy is one of the earliest disciplines of Botany, which can lay claims to being the oldest, the most basic and the most all-embracing of the biological sciences. Over 300000 different kinds of plants have been cataloged or are known. Scientists who study plants (botanists) like to sort them into categories (classify them). Without a way to put plants into categories, botanists would feel like their field of science was incredibly disorganized. Hence, the goal of taxonomy is to describe diversity, provide an insight to the evolutionary history (phylogeny), help to determine organisms (diagnostics) and allow for taxonomic estimations.

2.1 Botanical nomenclature

What is its name? Plant names not only contain valuable information about the plant, but also give the history of the plant, as well as its potential uses. Before the age of Linnaeus (林奈) in the 1700s, plants were typically referred to by common names, i. e. , the names that most people used for plants. For example, Basil, Dandelion, and Rose are all common names for plants, and are relatively straightforward. When one name applies to many plants, the situation becomes as twisted as a tangled vine. Like, what exactly does 'clover' refer to? Oxalis species or Trifolium species are good bets, but not sure bets. Fleabane could refer to many species of asters, which are good insect repellants. Plantain could refer to a small broadleaf weed, or to the edible (and may we add, delicious?) banana cousin.

Therefore, scientific names have several major advantages over common names and it should become apparent that common names, at best, will never be more than an unnecessary accessory to the framework provided by the formal system of naming organisms (nomenclature 命名). There are several minor advantages to common names. They are simple and hence easy to remember, usually descriptive of the plant, and larger numbers of people will sometimes know what you are talking about when common names are used than is the case with scientific names. However, the problems with common names far outweigh any advantages. Some of the most obvious problems with common names are: (i) there are over 300000 species of vascular plants and only a small percentage have common names; (ii) the same name is often used for different plants; (iii) common names are always in the local language, which prevents communication of plant identities between users of different languages; (iv) there is no formal process for the application of common names, it is usually not possible to determine when a common name was first used and the identity of the plant or plants to which it was applied; (v) the same plant may have different common names in different regions. The formal system of nomenclature or "scientific names" deals with these and other problems very effectively. Therefore, nomenclature is the giving and using of names. We give names as a code for something more important, and it codes things that we will want to remember and to use again. The name is the code that allows us into literature. The binomial system established by Linnaeus was universally accepted since 18th century in plant kingdom.

2.1.1 Plant taxonomy hierarchy

Once an organism was named, it has to be assigned an appropriate position in a systematic frame work of classification. This frame work is called taxonomic hierarchy by which the taxonomic groups are arranged in a definite order from higher to lower categories. Each category in the hierarchy is considered as a taxonomic unit and is known as taxon.

The species is the unit of plant classification. All individual plants, which resemble one another in important vegetative and reproductive characters and differ visibly from other plants, constitute a species. All the plants of a species are thus supposed to have descended from the same ancestor. The major categories (ranks) of the botanical classification system still in wide use today are, in descending order: Kingdom→Division (or Phylum) →Class→Order→Family→Genus→Species (Figure 2-1). Each organism can be placed into such a hierarchical system, a case as rose below. Biological systematists attempt to create classifications that reflect phylogeny; that is, a group of closely related species will be classified into a genus; closely related genera are placed in a family, and so on.

Kingdom	Plantae-Plants
Divisison	Magnoliophyta-Flowering plants
Class	Magnoliopsida-Dicotyledons
Order	Rosales
Family	Rosaceae
Genus	*Rosa*
Species	*Rosa chinensis* Jacq.

Figure 2-1 The hierarchy of taxonomic ranks.

2.1.2 Scientific names (Binomial)

At the simplest level of scientific classification, each plant has a name made up of two parts, a generic (or genus) name and a specific name or epithet. Together, these two names are referred to as a binomial (双名法), which is initiated by Linnaeus (Figure 2-2), a botanist from Sweden, also known as the 'father of modern taxonomy'.

A generic name is a 'collective name' for a group of plants. It indicates a grouping of organisms that all share a suite of similar characters. Ideally these should all have evolved from one common ancestor. The specific name, allows us to distinguish between different organisms within a genus.

Figure 2-2　Carolus Linnaeus

Binomial names are always written with the generic name first, starting with a capital letter, e.g.: *Ginkgo*; the specific epithet always follows the generic name, starting with a lower-case letter, e.g.: *biloba*; sometimes, nominator's name is placed at the end. The full species name or binomial is *Ginkgo biloba* L.

Generic and specific names are generally in Latin or are Latinised words from other languages, particularly Greek. Other derivations are also sometimes used, such as Aboriginal names or even acronyms. Specific epithets also need to conform to certain grammatical rules depending on the form of the generic name.

There are a number of levels of classification below that of species, with the most commonly used being subspecies and variety, abbreviated to 'subsp.', and 'var.' respectively. This allows further subdivision of plant groups to reflect the variation in form and distribution we see in nature. Meanwhile, the name of cultivar is added in the end of original species by single quotes.

In order to standardize the plant nomenclature in the world, botanical nomenclature is governed by the *International Code of Nomenclature for algae, fungi, and plants* (ICN), which replaces the *International Code of Botanical Nomenclature* (ICBN). Particularly, within the limits set by that code there is another set of rules, the *International Code of Nomenclature for Cultivated Plants* (ICNCP) which applies to plant cultivars that have been deliberately altered or selected by humans.

2.2　Phylogenetic system of classification

Linnaeus and the botanists before him tried to classify the plant kingdom using a single or a few characters chosen arbitrarily. They only thought about the convenience of following a system of classification solely to identify a particular plant. Such systems are, therefore, called artificial systems of classification. Later, plant-taxonomists conceived the idea that the plants belonged to some natural groups and they tried to designate and distinguish such groups and tried to classify the plant kingdom accordingly. Such systems are known as natural systems of classification.

Such natural grouping gives the idea that the individuals under one particular group are closely related to one another, although they believed in the fixity of species and simultaneous creation of all the groups by God. After the publication of the theory of '*Organic Evolution*' by Charles Darwin (查理·达尔文) and Alfred Wallace (阿尔弗雷德·华莱士) the taxonomist began to think about the origin of each of the natural groups from a more primitive group, or from an individual of a more primitive group. In other words, some of the natural groups are more primitive and some are more recent or advanced and the recent groups have been derived from some comparatively primitive group. A system of classification based on the idea of organic evolution attempting to find out the relation between the different groups, i.e., to trace the phylogeny of the groups, is called the phylogenetic system of classification. Some famous plant classifications as following:

2.2.1 Engler system

The first complete plant classification system, also widely accepted, in the world was established by Adolf Engler (阿道夫·恩格勒), who was a professor of botany from University of Berlin. In 1892, he published a system of classification mainly based on August Wilhelm Eichler (奥古斯特·威廉·艾希勒) in the book '*Syllabus der Vorlesungen*' as a guide to study the plants available in the Breslau Botanic Garden (Figure 2-3).

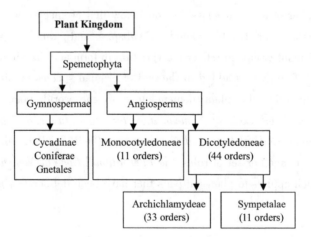

Figure 2-3 Outline of Engler system

The system of Engler has been widely used in the American and European continents. Engler divided the plant kingdom into thir-teen (13) Divisions. The thirteenth (13) Division is the Embryophyta Siphonogama (the seed-bearing plants i.e., Spermatophyta). It is divided into two Subdivisions, Gymnospermae and Angiospermae. The Angiospermae is divided into two Classes—

Monocotyledonae and Dicotyledonae. The Class Monocotyledonae is divided directly into 11 Orders.

In this system, Engler considered that in Embryophyta Siphonogama the flower without perianth is the primitive one. Thus, plants like Oak, Willow etc., with woody stem and unisexual apetalous flowers (Amentiferae), are treated as primitive Dicotyledons.

2.2.2 Hutchinson system

John Hutchinson (约翰·哈钦松) was a British botanist associated with Royal Botanic Gardens, Kew, England. He developed and proposed his system based on Bentham (边沁) and Hooker (虎克) and also on Bessey (贝斯). His phylogenetic system first appeared as *The Families of Flowering Plants* in two volumes, which contains Dicotyledons and Monocotyledons in first volume and second volume, respectively. However, the final revision of *The Families of Flowering Plants* was published just before his death on 2nd September 1972 and the 3rd i. e., the final edition, was published in 1973. A key feature of his third edition was based on the habit of the plant namely that herbaceous plants or Herbaceae are phylogenetically more recent than woody plants or Lignosae (Figure 2-4).

He divided the Phylum Angiospermae into two Subphyla Dicotyledones and Monocotyledones. The Dicotyledones are further divided into two divisions – Lignosae (arboreal) and Herbaceae (herbaceous). The Lignosae includes, fundamentally, the woody representatives derived from Magnoliales and Herbaceae includes most of the predominantly herbaceous families derived from Ranales. The subphylum Monocotyledones are divided into three divisions – Calyciferae, Corolliferae and Glumiflorae. So in the latest system of Hutchinson, the Dicotyledones consists of 83 orders and 349 families and Monocotyledones consists of 29 orders and 69 families.

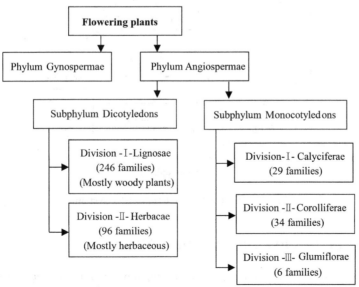

Figure 2-4 Outline of Hutchinson system

2.2.3 Cronquist system

The Cronquist system is a taxonomic classification system of flowering plants. It was developed by Arthur Cronquist (亚瑟·克朗奎斯特) in his texts '*An Integrated System of Classification of Flowering Plants*' (1981) and '*The Evolution and Classification of Flowering Plants*' (1968; 2nd edition, 1988). According to him 'many of the evolutionary trends bear little apparent relation to survival value and that there are some reversals'. In 1981, the system as laid out in '*Cronquist's An Integrated System of Classification of Flowering Plants*'. He divided the Division Magnoliophyta (Angiosperms) into two classes Magnoliatae (Dicotyledons) and Liliatae (Monocotyledons). He divided Magnoliatae into 6 subclasses and 55 orders, of which magnoliales is the primitive and Asterales is the advanced taxa. On the other hand, the class Liliatae has been divided into 4 subclasses and 18 orders, of which Alismatales is the primitive and Orchidales is the advanced taxa. The class Magnoliatae consists of 291 families and Liliatae with 61 families (Figure 2-5).

In this system, he discussed a wide range of characteristics important to phylogenetic system. He also provided synoptic keys designed to bring the taxa in an appropriate alignment. He also represented his classification in charts to show the relationships of the orders within the various subclasses. His system is more or less parallel to Takhtajan's system (塔赫他间系统), but differs in details. He considered that the Pteridosperms i. e., the seed ferns as probable ancestors of angiosperm.

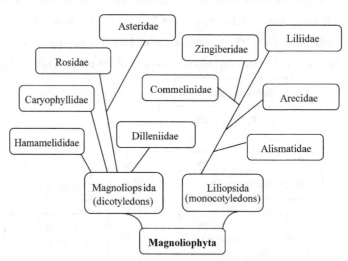

Figure 2-5　Outline of Cronquist's system

2.2.4 Angiosperm Phylogeny Group (APG)

The Angiosperm Phylogeny Group (APG) is an informal international group of systematic botanists who collaborate to establish a consensus on the taxonomy of flowering plants (angiosperms) that reflects new knowledge about plant relationships discovered through phylogenetic studies. An important motivation for the group was what they considered deficiencies in prior angiosperm classifications since they were not based on monophyletic groups (i. e. ,groups that include all the descendants of a common ancestor).

In the past, classification systems were typically produced by an individual botanist or by a small group. The result was a large number of systems. Different systems and their updates were generally favoured in different countries. Examples are the Engler system in continental Europe, the Bentham & Hooker system in Britain (particularly influential because it was used by Kew), the Takhtajan system in the former Soviet Union and countries within its sphere of influence and theCronquist system in the United States.

Before the availability of genetic evidence, the classification of angiosperms was based on their morphology (particularly of their flower) and biochemistry (the kinds of chemical compounds in the plant). After the 1980s, detailed genetic evidence analysed by phylogenetic methods became available and while confirming or clarifying some relationships in existing classification systems, it radically changed others. This genetic evidence created a rapid increase in knowledge that led to many proposed changes; stability was "rudely shattered". The impetus came from a major molecular study published in 1993 based on 5000 flowering plants and a photosynthesis gene (rbcL). This produced a number of surprising results in terms of the relationships between groupings of plants, for instance, the dicotyledons were not supported as a distinct group. At first there was a reluctance to develop a new system based entirely on a single gene. However, subsequent work continued to support these findings. These research studies involved an unprecedented collaboration between a very large number of scientists. Therefore, rather than naming all the individual contributors a decision was made to adopt the name Angiosperm Phylogeny Group classification, or APG for short (Figure 2-6).

The principles of the APG's approach to classification were set out as follows: (i) The Linnean's system of orders and families should be retained. (ii) Groups should be monophyletic (i. e. consist of all descendants of a common ancestor). (iii) A broad approach is taken to defining the limits of groups such as orders and families. (iv) Above or parallel to the level of orders and families, the term clades are used more freely.

Major achievements of the APG system were summarize as follows: (i) testing the repeatability and predictability of the APG system for angiosperms; (ii) resolving the systematic positions of some segregate taxa which were not placed based on morphological characters; (iii) proving that it is not reasonable to first divide angiosperms based on cotyledon character; (iv) demonstrating the importance of tricolpate/tricolporate pollen and derivatives for angiosperm classification; (v) finding that the centrifugal development of stamens in polyandrous groups have evolved independently many times and should not be used to delimit class or subclass of angiosperms; (vi) supporting that most of the families delimited by broad morphological characters are natural; and (vii) separating some families which are traditionally regarded as natural.

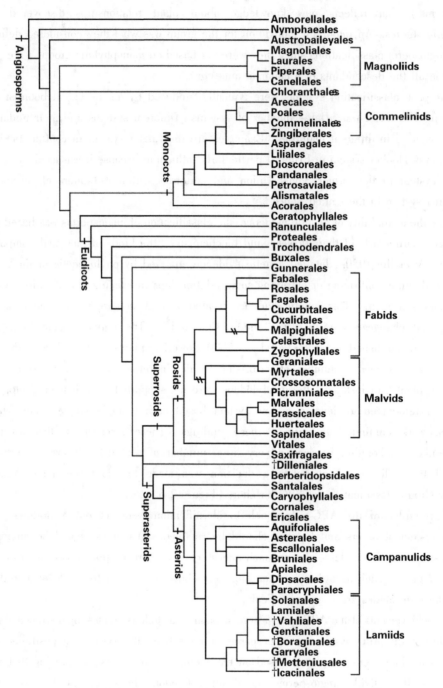

Figure 2-6 Outline of Angiospermic Phylogenetic Group IV[①]

---

① Chase M W, Christenhusz M J M, Fay M F, et al., 2016. An update of the Angiosperm Phylogeny Group classification for the orders and families of flowering plants: APG IV[J]. Botanical Journal of the Linnean Society, 181(1): 1-20.

However, the APG system also confronts some issues, such as: (i) how to harmonize the APG system and the morphology-based systems; (ii) establishing new morphological evolution theories on the basis of the APG system; (iii) determining whether it is enough to only use 'monophyly' as a criterion to circumscribe orders and families; (iv) determining morphological synapormorphies of those orders in the APG system; and (v) how to best compile a key to distinguish the orders and families of the APG system and to list their diagnostic characters for orders and families.

2.3 Taxonomic keys

A key is a device, which when properly constructed and used, enables a user to identify an organism. Keys are devices consisting of a series of contrasting or contradictory statements or propositions requiring the identifier to make comparisons and decisions based on statements in the key as related to the material to be identified. Thus, a taxonomic key is a device for quickly and easily identifying to which species an unknown plant belongs.

The key consists of a series of choices, based on observed features of the plant specimen. It provides a choice between two contradictory statements resulting in the acceptance of one and the rejection of the other. By making the correct choice at each level of the key, one can eventually arrive at the name of the unknown plant.

2.3.1 Types

Keys in which the choices allow only two (mutually exclusive) alternative couplets are known as dichotomous keys. In constructing a key, contrasting characters are chosen that divide the full set of possible species into smaller and smaller groups i.e. the statements typically begin with broad characteristics and become narrower as more choices are required.

Each time a choice is made, a number of species are eliminated from consideration and the range of possible species to which the unknown specimen may belong is narrowed. Eventually, after sufficient choices have been made, their range reduces to a single species and the identity of the unknown plant is revealed. Dichotomous comes from the Greek root dich meaning 'two' and temnein meaning 'to cut'. Couplets can be organized in several forms. The couplets can be presented using numbers (numeric) or using letters (alphabetical). The couplets can be presented together or grouped by relationships. There is no apparent uniformity in presentation of dichotomous keys.

There are two types of dichotomous keys. They differ in the method by which the couplets are organized and how the user is directed to successive choices.

(1) Indented Keys (also called yoked). Indents the choices (leads) of the couplet an equal distance from the left margin. The two choices of the couplet are usually labelled 1 and 1' or 1a and

1b. It is not necessary that the choices are numbered, but it helps. The user goes to the next indented couplet following the lead that was selected. For example:

  1a. Gymnosperm, ovule naked
    2a. Tree evergreen, leaves needlelike or linear
      3a. Leaves needlelike, short branchlets reduced ……………………………………
      ………………………………………………………… *Pinus massoniana* Lamb.（马尾松）
      3b. Leaves linear, without short branchlets ……… *Abies fabri*（Mast.）Craib（冷杉）
    2b. Tree deciduous, leaves flabellate …………………………… *Ginkgo biloba* L.（银杏）
  1b. Angiosperm, ovule enveloped by ovary
    4a. Shrub or small tree evergreen
    4b. shrub deciduous ……………………… *Chimonanthus praecox*（Linn.）Link（蜡梅）
      5a. Leaves lanceolate, follicle ……………………… *Nerium oleander* L.（欧洲夹竹桃）
      5b. Leaves elliptic, drupe ……………… *Osmanthus fragrans*（Thunb.）Lour.（木犀）

（2）Bracketed Keys. Provides both choices side-by-side. The choices of the couplet must be numbered (or lettered). It is very helpful if the previous couplet is given. This key has exactly the same choices as the first example. The choices are separated, but it is easy to see the relationships. While this key might be more difficult to construct, it gives more information to the user. For example:

  1a. Gymnosperm, ovule naked ……………………………………………………………… 2
  1b. Angiosperm, ovule enveloped by ovary ……………………………………………… 4
  2a. Tree evergreen, leaves needlelike or linear ………………………………………… 3
  2b. Tree deciduous, leaves flabellate …………………………… *Ginkgo biloba* L.（银杏）
  3a. Leaves needlelike, short branchlets reduced ……… *Pinus massoniana* Lamb.（马尾松）
  3b. Leaves linear, without short branchlets ……………… *Abies fabri*（Mast.）Craib（冷杉）
  4a. Shrub or small tree evergreen ………………………………………………………… 5
  4b. Shrub deciduous ……………………… *Chimonanthus praecox*（Linn.）Link（蜡梅）
  5a. Leaves lanceolate, follicle ……………………… *Nerium oleander* L.（欧洲夹竹桃）
  5b. Leaves elliptic, drupe ……………………… *Osmanthus fragrans*（Thunb.）Lour.（木犀）

2.3.2 Use and construct a taxonomic keys

（1）Main points for use of taxonomic keys

● Appropriate keys should be selected for the materials to be identified. The keys may be in a flora, manual, guide handbook, monograph, or revision. If the locality of an unknown plant is known, a flora, guide, or manual treating the plants of that geographic area may be selected.

● The introductory comments on format details, abbreviations, etc. should be read before using the key.

● Both the leads of a couplet should be read before making a choice. Even though the first lead may seem to describe the unknown material, the second lead may be even more appropriate.

● A glossary should be used to check the meaning of terms, which one does not understand.

● Several similar structures should be measured, when measurements are used in the key, e.g. several leaves and not a single leaf should be measured. No decision should be made on a single observation. Rather it is often desirable to examine several specimens.

● The results should be verified by reading a description, comparing the specimen with an illustration or an authentically named herbarium specimen.

(2) Main points for construction of taxonomic keys

● Constant characteristics rather than variable ones should be used.

● Proper measurements rather than terms like 'large' or 'small' should be used.

● Characteristics that are generally available to the user of the key rather than seasonal characteristics or those seen only in the field should be used.

● A positive choice should be made. The term 'is' instead of 'is not' should be used.

● If possible, one should start both choices of a pair with the same word.

● If possible, different pairs of choices should be started with different words.

● The descriptive terms should be preceded with the name of the part to which they apply.

References:

Shashank Goswami. Taxonomic Keys: Meaning, Suggestions and Types[EB/OL]. [2019-04-15]. http://www.biologydiscussion.com/plant-taxonomy/taxonomic-keys-meaning-suggestions-and-types/30278.

Centre for Australian National Biodiversity Researchand Australian National Herbarium. Plant Names[EB/OL]. (2017-5-9)[2019-04-15]. https://www.anbg.gov.au/cpbr/databases/names.html.

Plant Science 4 U. The Taxonomic Hierarchy of Plants (Mango)[EB/OL]. (2017-8)[2019-04-15]. https://www.plantscience4u.com/2017/08/the-taxonomic-hierarchy-of-plants-mango.html.

Stevens, P. F. (2001 onwards). Angiosperm Phylogeny Website. Version 14, July 2017. [EB/OL]. (2017-7)[2019-04-15]. http://www.mobot.org/MOBOT/research/APweb/.

Staffan Müller-Wille. Carolus Linnaeus-Swedish botanist[EB/OL]. (2019-9)[2019-04-15]. https://www.britannica.com/biography/Carolus-Linnaeus.

Bijoy G. Phylogenetic System of Plant Classification: Botany [EB/OL]. (2019-9) [2019-04-15]. http://www.biologydiscussion.com/plant-taxonomy/plant-classification/phylogenetic-system-of-plant-classification-botany/77691.

王伟,张晓霞,陈之端,等,2017. 被子植物APG分类系统评论[J]. 生物多样性, 25(4), 418-426.

谢春平,2014. 检索表教学与校园植物实践教学相结合的探索[J]. 职业教育研究 (5): 119-120.
王文采, Wen-Tsai W,1990. 当代四被子植物分类系统简介(一)[J]. 植物学报, 7(2): 1-17.
王文采,1990. 当代四被子植物分类系统简介(二)[J]. 植物学报, 7(3): 1-18.

Chapter 3
Gymnosperms 裸子植物

Gymnosperms are flowerless plants that produce cones and seeds. It is one group of seed plants that produce naked seeds not enclosed in an ovary and in some instances have motile spermatozoids. Gymnosperms were dominant in the Mesozoic Era, during which time some of the modern families originated (Pinaceae, Araucariaceae, Cupressaceae). Although since the Cretaceous Period gymnosperms have been gradually displaced by the more recently evolved angiosperms, they are still successful in many parts of the world and occupy large areas of Earth's surface. Conifer forests, for example, cover vast regions of northern temperate lands, and gymnosperms frequently grow in more northerly latitudes than do angiosperms.

Gymnosperms, though true seeds formed, the group differs from other group of seed-bearing plants (angiosperms), firstly in possessing naked ovules; secondly, in the lodging of pollen grains directly on the micropyle and thirdly, in the absence of true vessels and companion cells. This group is more ancient than angiosperms, claiming fossils as well as living members and form a bridge between the pteridophyta on one hand and the angiosperms on the other.

Generally, Chinese gymnosperm shows a decreasing trend of richness from south to north. The richness is high in mountainous areas and low in large plains and on large plateaus. From the species, genus to family levels, the area with high gymnosperm richness increases, and the center with high richness shifts southward. About 85% of all Chinese gymnosperm species are harbored in just 5% of Chinese land area. There are six richness centers of gymnosperm in China: (i) Eastern Himalaya-Hengduan Mountains-Qinling Mountains; (ii) joint area of Yunnan-Guizhou-Guangxi and the South Mountain Ranges; (iii) central China mountains; (iv) Huangshan-Wuyi Mountains; (v) southern mountains of Hainan Island; and (vi) central part of Changbai Mountains. However, Hengduan Mountains is a major variation center of gymnosperm in China.

3.1 Cycadaceae 苏铁科

Trees to shrubs, evergreen, mostly palmlike, trunk columnar. Bark usually thickened and roughened. Leaves borne at apex of trunk, 1(-3)-pinnately compound, spirally arranged. Two motile sperm cells.

There are 1 genus and 16 species (8 endemic) in China.

Cycas Linn. 苏铁属 I

Morphological characters are the same as those for the family.

Cycas revoluta Thunb. 苏铁 I (Figure 3-1)

Habit: trees, evergreen.

Bark: gray-black, scaly.

Leaf: 1-pinnately divided, glossy green, leaflets narrow with a sunken midrib, margins strongly recurved, apex acuminate.

Cone: pollen cones pale yellow, ovoid-cylindric; larg and erect microsporophylls narrowly

Figure 3-1 *Cycas revoluta* 苏铁

cuneate. Megasporophylls yellow to pale brown; sterile blade ovate to narrowly so, deeply laciniate, with numerous lobes; ovules with densely pale brown tomentose.

Seed: orange to red, obovoid or ellipsoid, somewhat compressed, sparsely hairy.

Distribution: Fujian, Guangdong and Taiwan; widely cultivated.

3.2 Ginkgoaceae 银杏科

Trees deciduous. Branchlets dimorphic: both long and short. Leaves sparsely and spirally arranged on long branchlets, fasciculate on short branchlets; flabellate, venation parallel, dichotomous. Seeds drupelike, succulent sarcotesta, a bony sclerotesta, and a membranous endotesta.

There is 1 genus and 1 species in China.

Ginkgo Linn. 银杏属

Morphological characters are the same as those for the family.

Ginkgo biloba Linn. 银杏 I (Figure 3-2)

Habit: trees, deciduous.

Figure 3-2 *Ginkgo biloba* 银杏

Bark: light gray or grayish brown, longitudinally fissured when old.

Twig: long branchlets pale brownish yellow initially, finally gray; short branchlets blackish gray, with dense, irregularly elliptic leaf scars.

Leaf: fan-shaped, blade pale green, turning bright yellow in autumn, 2 lobes in long branchlets, undulate distal and margin notched apex in short branchlets.

Cone: pollen cones ivory colored.

Seed: elliptic, narrowly obovoid, ovoid with rancid odor when ripe; sclerotesta white.

Distribution: Tianmu Mountains, widely cultivated.

3.3　Pinaceae 松科[①]

Trees evergreen or deciduous, rarely shrubs. Microsporophyll with 2 microsporangia; pollens often 2-saccate. Ovulate scales with 2 upright ovules adaxially; bracts free or adnate basally with seed scales.

There are 10 genera and 108 species in China.

1. *Keteleeria* Carr. 油杉属

Trees evergreen; short branchlets absent. Leaves linear to lanceolate, flattened, midvein raised on both sides. Seed cones terminal, solitary, erect, maturing in 1st year. Seed scales woody, persistent.

Keteleeria fortunei (Murr.) Carr. 油杉 (Figure 3-3)

Habit: trees, evergreen.

Bark: dark gray, rough, longitudinally fissured.

Twig: branchlets initially orange-red or reddish, turning yellowish gray or yellowish brown in 2nd or 3rd year, pubescent.

Leaf: linear, pectinately arranged in lateral sets, apex obtuse.

Cone: seed cones cylindric, green or pale green with slightly glaucous. seed scales compressed orbicular, rhombic-orbicular, exposed part glabrous abaxially, margin entire, apex convex, rounded.

Seed: seeds oblong; wing yellowish brown.

Distribution: Zhejiang, Fujian, Guangdong, Guangxi, Hunan, Jiangxi and Guizhou.

[①] Pinaceae belongs to a large group of trees known as conifers. The uses of the Pinaceae usually falls heavily on the wood and the wood by-products. Many wooden structures such as desks, chairs, tables, bedframes, house foundations, and even ships are made of pines, spruces, and cedars. Furthermore, wood by-products such as resin, pitch, and turpentine are important in wood-working.

Figure 3-3 *Keteleeria fortunei* 油杉

2. *Abies* Mill. 冷杉属

Trees evergreen; leaf scars orbicular on branchlets; short branchlets absent. Leaves spirally arranged, or pectinately arranged in lateral sets, linear, base twisted. Seed cones terminal, solitary, axillary; seed scales longer than bracts. Seeds wing well developed, persistent.

Abies firma Sieb. et Zucc. 日本冷杉 (Figure 3-4)

Habit: trees, evergreen.

Bark: black, rough, scaly, fissured.

Twig: branchlets grayish yellow, turning light gray or yellowish gray in 2nd or 3rd year, glabrous, or puberulent in groove; winter buds slightly resinous.

Leaf: linear, stomatal lines in 2 white bands abaxially, resin canals 2, median, apex emarginate or obtuse.

Cone: green, maturing yellow- or gray-brown, cylindric. Seed scales flabellate-trapeziform. Bracts often exserted, apex with abrupt, acute cusp.

Seed: cuneate-oblong wing.

Figure 3-4　*Abies firma* 日本冷杉

Distribution: native to Japan, cultivated in coastal areas of China.

3. *Pseudotsuga* Carr. 黄杉属 Ⅱ

Trees evergreen; short branchlets absent; winter buds ovoid, bright brown. Leaves spirally arranged, linear, resin canals 2. Bracts well developed, 3-lobed, straight or reflexed, with a cusp longer than lateral lobes. Seed cones terminal, solitary, Seeds small with wing.

Pseudotsuga sinensis Dode 黄杉 (Figure 3-5)

Habit: trees, evergreen.

Bark: gray or dark gray, irregularly and thickly scaly.

Twig: pale yellow or yellowish gray.

Leaf: linear, pectinately arranged, linear, apex emarginate slightly.

Cone: seed cones pale purple, glaucous, maturing purplish brown, ovoid to ellipsoid- or conical-ovoid. Bracts reflexed, cusp narrowly triangular, apex obtuse.

Seed: irregularly brown spotted abaxially, triangular-ovoid, slightly depressed; wing obliquely

Figure 3-5 *Pseudotsuga sinensis* 黄杉

ovate.

Distribution: southwest China and east China, endemic to China.

4. *Tsuga* (Endl.) Carr. 铁杉属

Trees evergreen; short branchlets absent; winter buds, not resinous. Leaves pectinately arranged; blade usually linear and flattened, resin canal 1 below vascular bundle. Seed cones terminal on 2nd year branchlets, solitary, pendulous, maturing in 1st year, small. Bracts included, rarely with slightly exserted.

Tsuga chinensis (Franch.) Pritz. 铁杉 (Figure 3-6)

Habit: trees, evergreen.

Bark: dark gray, longitudinally fissured, flaking.

Twig: branchlets pale yellow 1st year, slender, pubescent.

Leaf: linear, pectinately arranged, linear, short, margin entire, apex obtuse, entire or emarginate.

Cone: seed cones light green, maturing pale gray-yellow or pale brown, ovoid. Seed scales square-orbicular, pentagonal-ovate, or compressed orbicular, apex rounded. Bracts cuneate-obovate or obtriangular, apex erose.

Seed: small, with obliquely ovate wing.

Distribution: southwest China.

Figure 3-6 *Tsuga chinensis* 铁杉

5. *Cathaya* Chun et Kuang 银杉属

Trees evergreen; false short (lateral) branchlets. Leaves spirally arranged, linear-oblanceolate. Seed cones axillary.

Cathaya argyrophylla Chun et Kuang 银杉 I (Figure 3-7)

Habit: trees, evergreen.

Bark: dark gray, irregularly flaking.

Twig: branchlets yellow-brown, initially densely pubescent, turning dark yellow and glabrous; winter buds light yellow-brown.

Leaf: linear, dark green adaxially, clustered into a whorl on false short branchlets, puberulent, stomatal bands 2, abaxial, white, vascular bundle 1, resin canals 2, marginal, margin entire, apex rounded.

Cone: green, dark brown when mature, ovoid or ellipsoid. Seed scales suborbicular or compressed orbicular-ovate, densely pubescent on exposed part. Bracts 1/4-1/3 as long as seed scales.

Seed: obliquely ovoid; wing yellow-brown, obliquely ovate or elliptic-ovate.

Distribution: endemic to China, Guangxi, Sichuan, Hunan, and Guizhou.

Figure 3-7 *Cathaya argyrophylla* 银杉

6. *Picea* A. Dietr. 云杉属

Trees evergreen; pulvinus on long branchlets, short branchlets absent. Leaves included, linear, straight or curved, quadrangular, broadly rhombic, or flattened in cross section, stomatal lines adaxial or on each surface. Seed cones solitary, pendulous at maturity. Bracts included, small.

Picea asperata Mast. 云杉 (Figure 3-8)

Habit: trees, evergreen.

Bark: grayish brown, furrowed into irregular, rough, scaly plates.

Twig: branchlets included, glaucous; winter buds conical, scales appressed or slightly recurved in apical buds, recurved at base of branchlets, keeled.

Leaf: glaucous or not, linear, slightly curved, quadrangular-rhombic in cross section, stomatal lines 4-8 along each surface, apex acute.

Cone: green, maturing pale brown or reddish brown, cylindric-oblong. Seed scales at middle of cones obovate, margin entire or denticulate, apex rarely 2-lobed.

Seed: obovoid; wing pale brown, obovate-oblong.

Distribution: Gansu, Ningxia, Qinghai, Shaanxi, and Sichuan.

Figure 3-8 *Picea asperata* 云杉

7. *Larix* Mill. 落叶松属

Trees deciduous; branchlets strongly dimorphic: long branchlets and short branchlets. Leaves spirally arranged on long branchlets, in dense clusters on short branchlets. Cones borne at apex of short branchlets, solitary. Seed scales thin, leathery, persistent.

Larix kaempferi (Lamb.) Carr. 日本落叶松 (Figure 3-9)

Habit: trees, deciduous.

Bark: scaly, pale grey-brown.

Twig: long branchlets light yellow or light reddish brown, glaucous; short branchlets bearing rings of scale remnants.

Leaf: linear-oblanceolate, inconspicuously keeled abaxially, apex obtuse.

Cone: terminal, finally gray-brown, globose, rounded scales being markedly curved back when ripe.

Seed: brownish white mottled with red, ovoid-cuneate, slightly flattened; wing reddish yellow.

Distribution: native to Japan, but cultivated in north China and northeast China.

Figure 3-9 *Larix kaempferi* 日本落叶松

8. *Pseudolarix* Gord. 金钱松属

Trees deciduous; branchlets strongly dimorphic: long branchlets and short branchlets. Leaves spirally arranged on long branchlets, in dense clusters on short branchlets. Pollen cones terminal on short branchlets, cluster; seed cones solitary. Seed scales thick, woody, deciduous at maturity.

Pseudolarix amabilis (J. Nelson) Rehd. 金钱松 Ⅱ (Figure 3-10)

Habit: trees, deciduous.

Bark: gray-brown, rough, scaly, flaking.

Twig: long branchlets initially reddish brown, finally gray; short branchlets slow growing, bearing dense rings of leaf cushions; winter buds ovoid, scales free at apex.

Leaf: linear-needlelike, flexible, slightly curved or straight, apex acute.

Cone: green or yellow-green, maturing reddish brown, obovoid. Seed scales ovate-lanceolate. Bracts ovate-lanceolate, 1/4–1/3 as long as seed scales, margin denticulate.

Seed: seeds white, ovoid; wing light yellow.

Distribution: Fujian, Anhui, Hunan, Jiangxi, Zhejiang and Jiangsu.

Figure 3-10　*Pseudolarix amabilis* 金钱松

9. *Cedrus* Trew 雪松属

Trees evergreen; branchlets strongly dimorphic: long branchlets and short branchlets; winter buds small, scales persistent. Cones borne on apex of short branchlets, solitary, erect. Seed cones erect, maturing in 2nd (or 3rd) year; seed scales woody, deciduous at maturity.

Cedrus deodara (Roxb.) G. Don 雪松 (Figure 3-11)

Habit: trees, evergreen.

Bark: dark gray, cracking into irregular scales.

Twig: long branchlets pale grayish yellow, densely pubescent with some white powder in 1st year, after turning grayish.

Leaf: radially spreading on long branchlets, fascicled on short branchlets, needlelike, hard.

Cone: pale green, initially with some white powder, becoming reddish brown when ripe, ovoid. Seed scales flabellate-obtriangular.

Seed: triangular, winged.

Distribution: west of the Himalayas, cultivated widely as an ornamental in central China and east China.

Figure 3-11 *Cedrus deodara* 雪松

10. *Pinus* Linn. 松属[①]

Trees or rarely shrubs, evergreen; branchlets strongly dimorphic: long branchlets bearing scalelike leaves and spreading leaf bundles; short branchlets bearing leaves in bundles of 2-3-5. Leaves needlelike, base enclosed by membranous sheath. Seed scales woody, persistent; exposed

---

① *Pinus* is the largest genus of conifers, and it is nearly exclusively found in the northern hemisphere. *Pinus* is the most commercially important group of conifers, with wood, pulp, tar, and turpentine as its main products.

apex thickened and ridged (the apophysis), with a prominent protuberance (umbo).

Pinus bungeana Zucc. ex Endl. 白皮松 (Figure 3-12)

Habit: trees, evergreen.

Bark: bark irregularly flaking, inner bark pale, exfoliating in irregular, thin, scaly patches.

Twig: young branchlets gray-green, glabrous; winter buds red-brown, not resinous.

Leaf: 3-Needles per bundle, stiff, 1 vascular bundle, base with sheath shed.

Cone: solitary, often pale green, yellowish brown at maturity. Seed scales broadly oblong-cuneate, apex thickened; apophyses subrhombic; umbo dorsal, triangular, protruding.

Seed: gray-brown, subobovoid; wing loosely.

Distribution: Gansu, Henan, Hubei, Shaanxi, Shandong, Shanxi, and Sichuan.

Figure 3-12　*Pinus bungeana* 白皮松

Pinus massoniana Lamb. 马尾松 (Figure 3-13)

Habit: trees, evergreen.

Bark: red-brown toward apex of trunk, gray- or red-brown toward base, irregularly scaly and flaking.

Twig: 1st year branchlets yellowish brown; winter buds brown, ovoid-cylindric or cylindric.

Leaf: 2-needles per bundle, 4-8 resin canals, marginal, base with persistent sheath.

Cone: green, turning chestnut brown at maturity, ovoid, or ovoid-cylindric. Seed scales included, apophyses rhombic, slightly transversely ridged; umbo flattened, slightly sunken.

Seed: narrowly ovoid, winged.

Distribution: south of the Qinling Mountains-Huaihe River and Taiwan.

Figure 3-13 *Pinus massoniana* 马尾松

Pinus parviflora Sieb. et Zucc. 日本五针松 (Figure 3-14)

Habit: trees, evergreen.

Bark: pale gray, aging dull gray, furrowed with age into scaly plates.

Twig: 1st year branchlets yellow-brown with pubescent; winter buds ovoid, not resinous.

Leaf: 5-needles per bundle, short, stomatal lines white, 1 vascular bundle, base with sheath shed.

Cone: subsessile, ovoid or ovoid-ellipsoid, small.

Seed: nearly brown, irregularly obovoid, winged.

Figure 3-14 *Pinus parviflora* 日本五针松

Distribution: native to Japan, widely cultivated in many cities of China.

3.4 Taxodiaceae 杉科[①]

Trees; trunk straight. Leaves spirally arranged or scattered (decussate in *Metasequoia*). Microsporophylls and cone scales spirally arranged (decussate in *Metasequoia*). Bracts and cone scales usually spirally aranged (decussate in *Metasequoia*), sessile, opening when ripe (falling in *Taxodium*), semiconnate and free only at apex, or completely united; bracts occasionally rudimentary (in *Taiwania*); cone scales of mature cones flattened or shield-shaped, woody or leathery, 2-9-seeded on abaxial side.

There are 8 genera and 9 species in China.

---

① The Taxodiaceae genera, with the exception of Sciadopitys, are phylogenetically part of the Cupressaceae family, according to research. However, in order to comply with Chinese conservation rules, we continue to treat it as a separate family.

1. *Cunninghamia* R. Brown ex Rich. et A. Rich. 杉木属

Trees evergreen, monoecious. Leaves lanceolate, margin serrulate. Bracts persistent, large, leathery, margin irregularly and finely serrulate; seed scales minute with 3 seeds.

***Cunninghamia lanceolate* (Lamb.) Hook. 杉木 (Figure 3-15)**

Habit: trees, evergreen.

Bark: dark gray, longitudinally fissured, cracking into irregular flakes.

Twig: branches whorled or irregularly; winter buds ovoid.

Leaf: glossy deep green adaxially, narrowly linear - lanceolate, margin denticulate, apex usually symmetric and spinescent, spine.

Cone: pollen cone fascicles terminal. Seed cones terminal, 1-4 together, ovoid or subglobose; bracts glaucous, broadly ovate or triangular-ovate, irregular serrulate.

Seed: dark brown, oblong or narrowly ovate, narrowly winged laterally.

Distribution: while the exact native range is unknown due to extensive planting, it can be found almost to the south of the Qinling Mountains-Huaihe River.

Figure 3-15 *Cunninghamia lanceolate* 杉木

2. *Cryptomeria* D. Don 柳杉属

Trees, evergreen, monoecious. Leaves spirally 5-ranked, spreading or directed forward, subulate. Bracts and cone scales connate; cone scales thickened distally, woody, umbo with a central spine and 4 or 5(-7) toothlike projections on distal margin.

Cryptomeria japonica (Thunb. ex L. f.) D. Don 日本柳杉 (Figure 3-16)

Habit: trees, evergreen.

Bark: reddish brown, fibrous, peeling off in strips.

Twig: 1st year green, usually pendulous.

Leaf: spirally 5-ranked, spreading or directed forward, subulate to linear, straight or strongly incurved, rigid, stomatal bands with 2-8 rows of stomata on each surface.

Cone: pollen cones crowned into a raceme, yellow. Seed cones green, borne in groups, terminal; cone scales spirally arranged, apex usually recurved, umbo rhombic, distally with 4 or 5 (-7) toothlike projections.

Seed: 2-5 per cone scale, brown or dark brown, irregularly ellipsoid, winged.

Distribution: China and Japan. Usually cultivated as an ornamental and planted for timber.

Figure 3-16 *Cryptomeria japonica* 日本柳杉

3. *Taxodium* Rich. 落羽杉属

Trees deciduous or semievergreen, monoecious. Lateral branches alternate and deciduous in winter. 2 Ovules per bract axil; cone scales shield-shaped, woody at maturity, apex irregularly quadrangular. Seeds irregularly triquetrous.

Taxodium distichum (L.) Rich. 落羽杉 (Figure 3-17)

Habit: trees deciduous; pneumatophores present or absent around trunk.

Bark: brown, or gray, peeling off in long strips.

Twig: main branches spreading horizontally or ascending; lateral branchlets 2-ranked.

Leaf: linear and flat or subulate, 2-ranked on annual branchlets or not, turning dark reddish brown in fall; apex acute or sharply pointed.

Cone: pollen cones borne in terminal, racemes or panicles. Seed cones brownish yellow or white powdery, not glaucous, globose; cone scales shield-shaped, woody, obviously longitudinally grooved at apex.

Seed: reddish brown, irregularly triangular-pyramidal, sharply ridged.

Distribution: native to United States, commonly in many parts of China as an ornamental.

* *Taxodium distichum* var. *imbricatum* (Nutt.) Croom 池杉 is similar to the species, but the former's leaves on annual branchlets not 2-ranked, mostly subulate, a few linear and flat.

Figure 3-17 *Taxodium distichum* 落羽杉

4. *Metasequoia* Miki ex Hu et Cheng 水杉属

Leaves and cone scales decussate; lateral branchlets deciduous. Cone scales shieldlike, woody. Seeds compressed, winged all round.

***Metasequoia glyptostroboides* Hu et Cheng 水杉 Ⅰ (Figure 3-18)**

Habit: trees, deciduous, monoecious.

Bark: gray, or grayish brown; peeling off in long strips.

Twig: branchlets persistent or deciduous; winter buds ovoid, membranous, scales.

Leaf: deciduous together with lateral branchlet as a unit, decussate, 2 - ranked, spirally arranged; blade linear, flattened, soft.

Cone: pollen cones borne in spikes or panicles; microsporophylls decussate. Seed cones terminal or subterminal on previous year's growth, solitary; cone scales persistent, decussate, shieldlike, woody, grooved, base cuneate, distal part transversely rhombic.

Seed: compressed-obovoid, winged all round, apex emarginate.

Distribution: Hubei and Sichuan, cultivated widely in America and Europe.

Figure 3-18 *Metasequoia glyptostroboides* 水杉

3.5 Cupressaceae 柏科

Trees or shrubs evergreen. Leaves decussate or in whorls of 3, scalelike or needlelike. Seed scales flat or peltate, woody, leathery, or succulent, decussate, dehiscent or indehiscent when mature in 1st or 2nd (or 3rd) year; 1-many-seeded. Bracts almost completely enveloped by cone scales, free only at apex. Seeds winged or not.

There are 8 genera and 46 species in China.

1. *Platycladus* Spach 侧柏属

Leaves scalelike, 1-3mm. Branchlets arranged in a plane, spreading or ascending, flattened. Seed cones woody with 6 or 8 thick scales; bracts partly enveloped by cone scales, free apex a long, recurved cusp. Seeds wingless.

***Platycladus orientalis* (L.) Franco 侧柏 (Figure 3-19)**

  Habit: trees, evergreen.
  Bark: dark gray to light grayish brown, thin, flaking in long strips.
  Twig: branchlets arranged in a plane, spreading or ascending, flattened.
  Leaf: scalelike, decussate, 4-ranked, base decurrent, with an abaxial resin gland.

Figure 3-19 *Platycladus orientalis* 侧柏

Cone: pollen cones yellowish green, ovoid. Seed cones subovoid, solitary, dehiscent when mature in 1st year; cone scales 4 pairs, decussate, flat, thick, woody; free bract apex recurved cusp.

Seed: wingless or very narrow wing, ovoid or subellipsoid.

Distribution: almost all over China.

2. *Cupressus* Linn. 柏木属

Leaves scalelike, 1 – 3mm. Branchlets usually not arranged in a plane. Seed cones woody, maturing in 2nd year; fertile cone scales with 3 – numerous seeds. Seeds slightly flattened, narrow wings.

Cupressus funebris Endl. 柏木 (Figure 3-20)

Habit: trees, evergreen.

Bark: light grayish brown, flaking in long strips.

Twig: branchlets arranged in a plane, pendulous, green, slender, flattened.

Leaf: densely appressed, scalelike, apex sharply pointed; facial pairs with a linear abaxial gland.

Cone: pollen cones ellipsoid. Seed cones globose, ca. 1.0 cm in diam.; cone scales 4 pairs.

Seed: light brown, obovate-rhombic or suborbicular, flattened.

Distribution: native of central China, and now spread widely over China in cultivation.

Figure 3-20 *Cupressus funebris* 柏木

3. *Chamaecyparis* Spach 扁柏属

Branchlets arranged in a plane. Seed cones woody, maturing in 1st year; fertile cones scales with 2 seeds usually.

Chamaecyparis pisifera (Sieb. et Zucc.) Endl. 日本花柏（Figure 3-21）

Habit: trees, evergreen.

Bark: reddish brown, flaking in long strips.

Twig: branchlets pendulous slightly.

Leaf: scalelike, apex acute, white powder abaxially, facial leaves with an obscure abaxial gland.

Cone: seed cones dark brown, globose; cone scales 5-6 pairs, each fertile scale with 1 or 2 seeds.

Seed: narrowly obovoid to transversely ellipsoid, winged.

Distribution: native to Japan, cultivated for ornamental in China.

Figure 3-21 *Chamaecyparis pisifera* 日本花柏

4. *Fokienia* A. Henry et H. H. Thomas 福建柏属

Leaves 2–10 mm. Seed cones woody, dehiscent when mature in 2nd year. Seeds with 2 apical, unequal wings.

Fokienia hodginsii (Dunn) A. Henry et H. H. Thomas 福建柏 II (Figure 3-22)

Habit: trees, evergreen.

Bark: purplish brown, nearly smooth or irregularly fissured.

Twig: branchlets arranged in a plane, flattened.

Leaf: larger, ca. 6mm, decussate, scalelike, dark green adaxially, white powder abaxially.

Cone: pollen cones yellowish green. Seed cones brown, dehiscent when mature in 2nd year; cone scales 6–8 pairs, decussate, peltate, woody, fertile scales 2-ovulate; free bract apex a mucro.

Seed: ovoid, with a prominent umbilicus and 2 apical, unequal wings.

Distribution: east China and southwest China.

Figure 3-22 *Fokienia hodginsii* 福建柏

5. *Juniperus* Linn. 刺柏属

Seed cones succulent, indehiscent or slightly dehiscent when mature. Seeds wingless.

***Juniperus chinensis* Linn. 圆柏 (Figure 3-23)**

Habit: shrubs or trees, evergreen.

Bark: grayish brown, flaking in long strips.

Twig: branches spreading; branchlets straight or slightly curved, terete or 4-angled.

Leaf: needlelike, present on both young and adult plants, decussate or in whorls of 3, loosely arranged; scalelike, present on adult plants, decussate, closely appressed.

Cone: pollen cones yellow, ellipsoid. Seed cones brown when ripe, often glaucous, with whitepowder.

Seed: ovoid, flattened, apex blunt.

Distribution: almost all over China.

Figure 3-23 *Juniperus chinensis* 圆柏

Juniperus formosana Hayata 刺柏 (Figure 3-24)

Habit: shrubs or trees, evergreen.

Bark: brown, flaking in long strips.

Twig: branches spreading or ascending; branchlets pendulous, 3-angled.

Leaf: whorls of 3, linear-lanceolate or linear-needlelike, base jointed, not decurrent, apex sharply pointed.

Cone: pollen cones axillary, globose. Seed cones axillary, light reddish brown when ripe, glaucous, with 6 fused scales in 2 alternating whorls, often 3-seeded.

Seed: ovoid-triangular, 3- or 4-ridged, apex pointed.

Distribution: south of the Yangtze River, Taiwan, Qinghai and Gansu.

Figure 3-24 *Juniperus formosana* 刺柏

3.6 Podocarpaceae 罗汉松科

Microsporophylls numerous, spirally arranged; microsporangia 2; pollen 2-saccate. Ovule (inverted) or inclined. Seed drupelike or nutlike, wholly or partly enveloped in a sometimes colored and succulent epimatium.

There are 4 genera and 12 species in China.

Podocarpus L. Her. ex Pers. 罗汉松属 II

Leaves spirally arranged, linear, lanceolate, or ovate-elliptic, with a single, obvious, often raised midvein on both surfaces. Epimatium wholly enveloping seed, sometimes colored and succulent. Seed ripening in 1st year, drupelike, dry, or leathery.

Podocarpus macrophyllus (Thunb.) Sweet 罗汉松 II (Figure 3-25)

Habit: trees, evergreen.
Bark: gray or grayish brown, peeling off in thin flakes.

Figure 3-25 *Podocarpus macrophyllus* 罗汉松

Twig: glabrous or pubescent.

Leaf: spirally arranged, sessile; blade dark green adaxially, pale green abaxially, linear-lanceolate, midvein prominently raised adaxially, base cuneate, apex mucronate.

Cone: pollen cones axillary, usually in clusters on very short peduncle, spikelike. Seed-bearing structures axillary, solitary, pedunculate. Receptacle red or purplish red when ripe, columnar. Epimatium purplish black when ripe, with white powder.

Seed: ovoid.

Distribution: south of the Yangtze River.

 * *Podocarpus macrophyllus* var. *maki* sieb et zucc. 短叶罗汉松 II is similar to the species, but leaves denser, shorter, ca. 5 cm long, ca. 5mm wide, apex obtuse.

3.7　Cephalotaxaceae 三尖杉科

Bud scales persistent. Leaves 2-ranked, blade linear, linear-lanceolate, stomatal bands 2, white. Pollen nonsaccate. Seed cones borne from axils of terminal bud scales. Seeds drupelike, completely enclosed by succulent aril, ovoid, ellipsoid, or globose, apex mucronate.

There are 1 genus and 9 species in China.

Cephalotaxus Sieb. et Zucc. ex Endl. 三尖杉属

Morphological characters are the same as those of the family.

Cephalotaxus sinensis (Rehd. et E. H. Wils.) H. L. Li 粗榧 (Figure 3-26)

Habit: shrubs or small trees, evergreen.

Bark: gray, or grayish brown, with irregular cracking.

Twig: leafy branchlets elliptic, oblong.

Leaf: blade green adaxially, linear, straight, flat, ca. 2-5 cm long, leathery and soft, stomatal bands white, base cuneat, symmetric, margin narrowly revolute, apex acute and shortly mucronate to long acuminate.

Cone: pollen-cone capitula globose. Seed cones solitary or borne 6-7 together; seed scales grayish green, ovate, apex shortly cuspidate. Aril red or reddish purple when rip.

Seed: ovoid or obovoid to ellipsoid, apex mucronate or cuspidate.

Distribution: from eastern China to Sichuan and west Yunnan and south Henan.

 * *Cephalotaxus fortunei* Hook. f. 三尖杉 is similar to the species, but leaves ca. 4-13 cm long, ca. 3-4.5mm wide, flexible, base cuneate or shortly attenuate, asymmetric.

Figure 3-26 *Cephalotaxus sinensis* 粗榧

3.8 Taxaceae 红豆杉科

Trees or shrubs evergreen. Leaves spirally arranged or decussate, linear or lanceolate. Pollen is nonsaccate. Ovule solitary, borne at apex of floral axis, erect. Seed drupelike or nutlike, partiallyenclosed in a succulent, saccate or cupular aril, or completely enclosed within aril.

There are 4 genera and 11 species in China.

1. *Taxus* Linn. 红豆杉属 Ⅰ

Branchlets irregularly alternate. Leaves linear, with midvein prominent adaxially; abaxial stomatal bands of leaves pale yellow or pale grayish green. Aril red when ripe, cupular.

Taxus wallichiana Zucc. 喜马拉雅红豆杉 Ⅰ

Habit: trees or shrubs, evergreen.

Bark: variably colored, peeling off in strips or cracking and falling off as thin scales.

Twig: winter bud scales deciduous or some persistent at base of branchlets.

Leaf: subsessile, linear to lanceolate, midvein slightly elevated adaxially, densely and evenly papillate abaxially, stomatal bands pale yellowish, apex acuminate.

Cone: pollen cones scattered along 2nd year branchlet axis, pale yellowish, ovoid. Aril red or orange when ripe, often translucent.

Seed: ovoid or obovoid, slightly flattened, usually with obtuse ridges; apex with small mucro; hilum elliptic to suborbicular or rounded-trigonous.

Distribution: from southwest China to east China, and Taiwan. There are 2 varieties, including:

* *Taxus wallichiana* var. *chinensis* (Pilger) Florin 红豆杉 I: leaves linear, straight to distally falcate, thick textured, midvein of same color as stomatal band, densely and evenly papillate, margin flat in living state.

* *Taxus wallichiana* var. *mairei* (Lemée et H. Léveillé) L. K. Fu et Nan Li 南方红豆杉 I (Figure 3-27): leaves linear, usually falcate, thick, midvein of different color to stomatal band, not papillate, or with papillae scattered on midvein or in 1-several lateral rows adjacent to stomatal band, margin revolute.

Figure 3-27 *Taxus wallichiana* var. *mairei* 南方红豆杉

2. *Torreya* Arn. 榧树属 Ⅱ

Leaves with midvein inconspicuous adaxially. Pollen sacs borne on outer side of microsporophylls, with distinct adaxial and abaxial surfaces. Seed-bearing structures paired in leaf axils, sessile; seed completely enclosed within aril.

***Torreya grandis* Fort. et Lindl.** 榧树 Ⅱ (Figure 3-28)

Habit: trees, evergreen.

Bark: light yellowish gray, with irregular vertical fissures.

Twig: branchlets turning yellowish green after 1st year.

Leaf: linear-lanceolate, usually straight, with no conspicuous grooves, midvein indistinct adaxially, stomatal bands yellowish green, base obtuse or broadly rounded, apex symmetrically, cuspidate.

Cone: pollen cones columnar; bracts conspicuously ridged. Aril pale purplish brown and white powdery when ripe, apex obtuse-rounded or rounded and cuspidatae.

Seed: ellipsoid to ovoid.

Distribution: east China, Guizhou and Hunan.

Figure 3-28 *Torreya grandis* 榧树

References:

James W Hardin, Donald Joseph Leopold, Fred M White, 2001. Harlow & Harrar´s textbook of dendrology [M]. Boston: McGraw-Hill.

Wu Z. Y. & P. H. Raven, 1999. Flora of China (Vol. 4) [M]. Beijing: Science Press&St. Louis: Missouri Botanical Garden Press.

Simpson M G, 2010. Plant Systematics [M]. 2th edition. San Diego: Academic Press.

Kramer Karl Ulrich, Peter Shaw Green, 1990. Pteridophytes and gymnosperms [M]. Berlin, Heidelberg: Springer-Verlag.

祁承经,汤庚国,2005. 树木学:南方本[M]. 2版. 北京:中国林业出版社.

李果,沈泽昊,应俊生,等,2009. 中国裸子植物物种丰富度空间格局与多样性中心[J]. 生物多样性, 17(3): 272-279.

中国科学院中国植物志委员会,1978. 中国植物志:第七卷[M]. 北京:科学出版社.

Chapter 4

Angiosperms 被子植物

The angiosperms, or flowering plants, are one of the major groups of extant seed plants and arguably the most diverse major extant plant group on the planet, with at least 260000 living species classified in 453 families. They occupy every habitat on the Earth except extreme environments such as the highest mountaintops, the regions immediately surrounding the poles, and the deepest oceans. They live as epiphytes (i. e., living on other plants), as floating and rooted aquatics in both freshwater and marine habitats, and as terrestrial plants that vary tremendously in size, longevity, and overall form. They can be small herbs, parasitic plants, shrubs, lianas, or giant trees. There is a huge amount of diversity in chemistry (often as a defense against herbivores), reproductive morphology, and genome size and organization that is unparalleled in other members of the Plant Kingdom. Furthermore, angiosperms are crucial for human existence; the vast majority of the world's crops are angiosperms, as are most natural clothing fibers. Angiosperms are also sources for other important resources such as medicine and timber.

Despite their diversity, angiosperms are clearly united by a suite of synapomorphies, including: (i) ovules that are enclosed within a carpel, that is, a structure that is made up of an ovary, which encloses the ovules, and the stigma, a structure where pollen germination takes place; (ii) double fertilization, which leads to the formation of an endosperm (a nutritive tissue within the seed that feeds the developing plant embryo); (iii) stamens with two pairs of pollen sacs; (iv) features of gametophyte structure and development; and (v) phloem tissue composed of sieve tubes and companion cells.

Traditionally, the flowering plants are divided into two groups, (i) dicotyledoneae or magnoliopsida, and (ii) monocotyledoneae or liliopsida. Despite the problems in recognizing basal angiosperm taxa, the standard distinctions between dicots and monocots are still quite useful. It must be pointed out, however, that there are many exceptions to these characters in both groups, and that no single character in the list below will infallibly identify a flowering plant as a monocot or dicot. The table 4-1 summarizes the major morphological differences between monocots and dicots:

Table 4-1 Comparison of monocotyledoneae and dicotyledoneae

MONOCOTS	DICOTS
Embryo with single cotyledon	Embryo with two cotyledons
Pollen with single furrow or pore	Pollen with three furrows or pores
Flower parts in multiples of three	Flower parts in multiples of four or five
Major leaf veins parallel	Major leaf veins reticulated
Stem vascular bundles scattered	Stem vascular bundles in a ring
Roots are adventitious	Roots develop from radicle
Secondary growth absent	Secondary growth often present

4.1 Magnoliaceae 木兰科[①]

Trees or shrubs, evergreen or deciduous. Stipules 2, remaining annular scar on twig. Leaves simple, alternate, margin entire or rarely lobed. Flowers terminal or terminal on axillary brachyblasts, solitary. Tepals 6-9(-45), in 2 to many whorls, 3(-6) per whorl. Carpels and stamens many, spirally arranged on an elongated torus. Aggregate follicle or samaroid.

There are 13 genera and 112 species in China.

1. *Magnolia* Linn. 木兰属

Trees or shrubs, deciduous or evergreen. Flowers terminal on brachyblasts, solitary, bisexual; tepals 9-15, 3 per whorl, sometimes sepal-like. Gynoecium linked to androecium, without a gynophore; ovules 2(-4) per carpel. Fruit a follicle.

Magnolia denudate Desr. 玉兰 (Figure 4-1)

Habit: trees, deciduous.

Bark: deep gray, coarse and fissured.

Twig: grayish brown; winter buds densely pale grayish yellow long sericeous; stipular scar conspicuous in twig.

Leaf: obovate, broadly obovate, or obovate-elliptic, apex abrupt acute, base cuneate or broadly.

[①] Unlike most angiosperms, whose flower parts are in whorls (rings), the Magnoliaceae have their stamens and pistils in spirals on a conical receptacle. This arrangement is found in some fossil plants and is believed to be a basal or early condition for angiosperms. The flowers also have parts not distinctly differentiated into sepals and petals, while angiosperms that evolved later tend to have distinctly differentiated sepals and petals. The poorly differentiated perianth parts that occupy both positions are known as tepals.

Figure 4-1　*Magnolia denudata* 玉兰

　　Flower: large, appearing before leaves, fragrant. Tepals 9, white, oblong-obovate. Gynoecium pale green, glabrous.

　　Fruit: aggregate follicle, cylindric; mature carpels brown, thickly woody, white lenticellate.

　　Distribution: cultivated widely all over China.

***Magnolia grandiflora* Linn.** 荷花玉兰(广玉兰)(**Figure 4-2**)

　　Habit: trees evergreen.

　　Bark: pale brown to gray, thinly scaly fissured.

　　Twig: sturdy, with brown tomentose.

　　Leaf: elliptic to obovate-oblong, thickly leathery, adaxially deep green and glossy, abaxial surfaces densely brown to grayish brown shortly tomentose, base cuneate, apex obtuse to shortly mucronate.

　　Flower: fragrant. Tepals 9-12, white, obovate, thickly fleshy.

　　Fruit: aggregate follicle, densely brown to pale grayish yellow tomentose.

　　Distribution: native to southeast North America, cultivated in south of the Yangtze River

Figure 4-2　*Magnolia grandiflora* 荷花玉兰

basin in China.

***Magnolia officinalis* Rehd. et E. H. Wils. 厚朴 Ⅱ (Figure 4-3)**

　　Habit: trees, deciduous.

　　Bark: brown, thick, not fissured.

　　Twig: pale yellow to grayish yellow, sturdy; terminal buds large; stipular scar conspicuous in twig.

　　Leaf: oblong-obovate, nearly leathery, base cuneate, margin entire or slightly wavy, apex shortly acute, obtuse, emarginate, or sometimes 2-lobed.

　　Flower: white and large, fragrant. Tepals 9-12, outer 3 tepals pale green, tepals of inner 2 whorls obovate-spoon-shaped. Stamens numerous, filaments red.

　　Fruit: aggregate follicle, ellipsoid-ovoid.

　　Distribution: east China, central China, southwest China, Gansu, Shaanxi, and Hunan.

Figure 4-3 *Magnolia officinalis* 厚朴

Magnolia zenii Cheng 宝华玉兰 Ⅱ (Figure 4-4)

Habit: trees, deciduous.

Bark: grayish white, smooth.

Twig: purple, winter buds densely long sericeous; stipular scar conspicuous in twig.

Leaf: obovate-oblong to oblong, abaxially pale green, midvein with long curved trichomes.

Flower: appearing before leaves, fragrant. Tepals 9, nearly spoon-shaped, inner tepals white but outside pale purplish red from base to middle and apically white.

Fruit: aggregate follicle, cylindric.

Distribution: native to Baohua Shan Mountain of Jiangsu Province.

Figure 4-4 *Magnolia zenii* 宝华玉兰

Magnolia liliiflora Desr. 紫玉兰（Figure 4-5）

  Habit: shrubs, deciduous, clustered.

  Bark: grayish brown.

  Twig: greenish purple to pale purplish brown; stipular scar conspicuous in twig. Winter buds densely long sericeous.

  Leaf: elliptic-obovate to obovate, abaxially grayish green and pubescent along veins, base gradually narrowing along petiole to stipular scar, apex acute to acuminate.

  Flower: appearing at same time with leaves, slightly fragrant. Tepals 9-12, outer 3 tepals purplish green, sepal-like, caducous; tepals of inner 2 whorls purple to purplish red outside and whitish inside, petal-like, fleshy.

  Fruit: aggregate follicle, dark purplish brown, cylindric.

  Distribution: native to Chongqing, Fujian, Hubei, Shaanxi, Sichuan, and Yunnan; but cultivated in many regions.

Figure 4-5 *Magnolia liliiflora* 紫玉兰

Magnolia biondii Pampan. 望春玉兰 (Figure 4-6)

Habit: trees, deciduous.

Bark: pale gray, smooth.

Twig: grayish green, glabrous. Stipular scar conspicuous in twig.

Leaf: narrowly elliptic to ovate, base broadly cuneate to obtuse, apex acute to shortly acuminate.

Flower: appearing before leaves, fragrant. Tepals 9, outer 3 tepals sepaloid, purplish red; tepals of middle and inner whorls white but usually outside purplish red at base, spoon-shaped.

Fruit: aggregate follicle, cylindric, usually withered because of carpels partly undeveloped.

Distribution: native to Gansu, Shaanxi, Henan, Hubei and Sichuan.

Figure 4-6　*Magnolia biondii* 望春玉兰

2. *Michelia* Linn. 含笑属

　　Trees or shrubs, evergreen. Stipules hooded, annular scar persistent on petiole or twig. Flowers pseudoaxillary on a brachyblast, solitary, bisexual, usually fragrant. Tepals 6-many, 3 or 6 per whorl. Gynoecium with a gynophore; ovules 2 to several per carpel. Fruit follicle.

***Michelia figo* (Lour.) Spreng. 含笑 (Figure 4-7)**

　　Habit: shrubs or small trees, evergreen.

　　Bark: grayish brown.

　　Twig: young twigs, buds, petioles, and brachyblasts densely yellowish brown tomentose. Stipular scar conspicuous in twig.

　　Leaf: petiole short, narrowly elliptic to obovate-elliptic, base cuneate to broadly cuneate, apex obtusely acute, margin entire.

　　Flower: tepals 6, pale yellow but margin sometimes red to purple. Gynophore ca. 6 mm; gynoecium ca. 7 mm, exceeding androecium, glabrous.

　　Fruit: small, aggregate follicle.

Figure 4-7 *Michelia figo* 含笑

Distribution: native to south China, but cultivated in the Yangtze River basin commonly.

Michelia maudiae Dunn 深山含笑 (Figure 4-8)

Habit: trees, evergreen.

Bark: pale gray or grayish brown, thin.

Twig: young twigs, buds, abaxial surfaces, and bracts with white powdery. Stipular scar conspicuous in twig.

Leaf: oblong-elliptic, leathery, abaxially grayish green and glaucous, adaxially deep green and glossy, base cuneate, apex abruptly shortly acuminate and with an obtuse tip.

Flower: tepals 9, white. Filaments pale purple, flat. Gynophore 5-8 mm, carpels green, narrowly ovoid.

Fruit: aggregate follicle.

Distribution: east China, central China and southwest China.

Figure 4-8 *Michelia maudiae* 深山含笑

Michelia chapensis Dandy 乐昌含笑 (Figure 4-9)

Habit: trees, evergreen.

Bark: gray to dark brown.

Twig: glabrous or nodes grayish puberulous when young. Stipular scar conspicuous in twig.

Leaf: leaf blade obovate, thinly leathery, adaxially deep green and glossy, base cuneate, apex acute to shortly acuminate, acumen obtuse.

Flower: tepals 6, in 2 whorls, pale yellow. Gynophore ca. 7 – 10 mm, densely silvery appressed puberulous.

Fruit: aggregate follicle.

Distribution: Jiangxi, Hunan, Guangdong and Guangxi; also recorded in Vietnam.

Figure 4-9 *Michelia chapensis* 乐昌含笑

3. *Liriodendron* Linn. 鹅掌楸属

Trees, deciduous. Winter buds surrounded by 2 connate stipules. Leaf blade with 1 or 2 lateral lobes near base, apex truncate to emarginate. Flowers terminal, solitary, bisexual, not fragrant, appearing at same time as leaves. Tepals 9, in 3 whorls, subequal. Fruit fusiform, winged seeds.

***Liriodendron chinense* (Hemsl.) Sarg. 鹅掌楸 Ⅱ (Figure 4-10)**

Habit: trees, deciduous.

Bark: grayish white, ridged or plated.

Twig: gray to grayish brown. Stipular scar conspicuous in twig.

Leaf: membranous to papery, abaxially glaucous, base truncate to slightly cordate and with 1 lateral lobe near base of each side, apex 2-lobed.

Flower: cupular. Tepals 9, outer 3 tepals green, sepal-like; tepals of inner 2 whorls green with yellow striations, erect, petal-like.

Fruit: nutlet, aggregate of winged seeds.

Distribution: east China, central China and southwest China.

* The species is similar to *Liriodendron tulipifera* Linn. 北美鹅掌楸, but the latter 2-lobed both side of leaf blade and without powdery abaxially.

Figure 4-10 *Liriodendron chinense* 鹅掌楸

4.2 Illiciaceae 八角科[①]

Shrubs and small trees, evergreen. Leaves simple, alternate, glandular and fragrant. Flowers

---

① Sometimes, this family was placed in Magnoliaceae as one of genera, but we deal with it as an independent family in the textbook. Aromatic oils obtained from some members of this family are used for flavorings and as carminatives; oil derived from *Illicium anisatum* is poisonous. Chinese star anise, used widely for flavoring wine and cooking, is obtained from *I. verum*. The Chinese drug Bajiao huixiang (八角茴香), are used to treat vomiting, epigastric pain, and abdominal colic, is derived from ripe fruits of *I. verum*.

solitary. Tepals 2 or 3 rows, the inner ones like petals, but the outer ones smaller and like bracts. Stamens 4 to many. Carpels 5-21, in a whorl; ovary 1-loculed. Fruit a follicle, star-shaped, dehiscent.

There are 1 genus and 27 species in China.

Illicium Linn. 八角属

The morphological characteristics are the same as the family's.

Illicium lanceolatum A. C. Smith 红毒茴(莽草) (**Figure 4-11**)

Habit: trees, evergreen.

Bark: pale grey to greyish brown.

Twig: slender.

Leaf: alternate, or more densely placed at branchlet ends, leathery, lanceolate, oblanceolate or obovate-elliptic, base narrowly cuneate, apex caudate; reticulate veins inconspicuous.

Flower: axillary or subterminal, solitary or in fascicles of 2 or 3; perianth segments 10-15, fleshy, red to dark red; carpels 10-14; stamens 6-11.

Fruit: follicles 10-14, apex with a recurved 3-7 mm hooked beak.

Distribution: east China, central China and Guizhou.

* the species is similar to *Illicium verum* J. D. Hooker 八角, but the latter carpels 4-10, fruit with 2-10 follicles. *I. verum* is a traditional spice in Chinese food.

Figure 4-11 *Illicium lanceolatum* 红毒茴

4.3　Lauraceae 樟科[①]

Bark and foliage usually aromatic. Leaves simple, estipulate. Inflorescence panicle, spike, or, racemes. Flowers small, greenish, yellowish, or white. Perianth 4 or 6. Androecium typically of 4 whorls of 3 stamens each; anthers basifixed, 2-celled or 4-celled at anthesis, inner 3rd whorl extrorse, dehiscing by flaplike valves opening upward. Fruit a drupe or berry.

There are 25 genera and 445 species in China.

1. *Cinnamomum* Schaeff. 樟属

Evergreen trees or shrubs. Branchlets, and leaves strong scented. Leaves alternate, subopposite, or opposite, leathery. Flowers bisexuality, panicle axillary. Fruit fleshy, subtended by a perianth cup; perianth cup cupuliform.

Cinnamomum camphora (Linn.) Presl 樟 (Figure 4-12)

Habit: trees, evergreen and large.

Bark: yellow-brown, irregularly and longitudinally fissured.

Twig: brownish, terete, glabrous. Terminal buds broadly ovoid.

Leaf: alternate; ovate to oblong-ovate, powdery abaxially, triplinerved or sometimes inconspicuously 5-nerved, base broadly cuneate, margin entire and undulate, apex acute.

Flower: inflorescence a panicle axillary. Flowers small, yellowish green.

Fruit: drupe, purple-black, ovoid or subglobose; perianth cup in fruit cupuliform.

Distribution: south of the Yangtze River in mainland and Taiwan province.

Cinnamomum japonicum Sieb. 天竺桂 Ⅱ (Figure 4-13)

Habit: trees, evergreen.

Bark: dark gray.

Twig: red or red-brown, slender, terete, glabrous, scented.

Leaf: subopposite or those on upper part of branchlet alternate, ovate-oblong or oblong-lanceolate, leathery, glabrous on both surfaces, triplinerved, base broadly cuneate or obtuse, margin entire, apex acute.

Flower: inflorescence a panicle axillary. Perianth lobes 6, ovate.

Fruit: oblong, glabrous; perianth cup in fruit shallowly cupuliform.

[①] The laurel or avocado family (Lauraceae) is among the most frequent and ecologically most important woody plant families in moist tropical and subtropical forests worldwide. Because the family is so ancient and was so widely distributed on the Gondwana supercontinent, modern species commonly occur in relict populations isolated by geographical barriers, for instance on islands or tropical mountains. Some important economic plants, including fruit from *Persea americana* (Avocado pear, 鳄梨), cinnamon and camphor from *Cinnamomum* spp., aromatic oils oils from *Lindera* (benzoin) and *Sassafras*, and fragrant woods used in cabinet-making.

Figure 4-12 *Cinnamomum camphora* 樟

Figure 4-13 *Cinnamomum japonicum* 天竺桂

Distribution: east China.

2. *Phoebe* Nees 楠属

Evergreen trees or shrubs. Leaves alternate, pinnately veined. Flowers bisexual, cymose paniculate or subracemose. Perianth lobes 6, erected. Fruit a drupe, ovoid to globose, base surrounded by persistent and enlarged perianth lobes.

Phoebe sheareri (**Hemsl.**) **Gamble** 紫楠 (**Figure 4-14**)

Habit: trees, evergreen.

Bark: gray-white.

Twig: branchlets, petioles, and inflorescences densely yellowish brown or gray-blackish pubescent to tomentose.

Leaf: obovate to broadly oblanceolate, leathery, abaxially pubescent to villous, densely reticulate raised abaxially.

Flower: inflorescence a panicle, brown hair conspicuous.

Fruit: drupe, ovoid, persistent perianth lobes hairy on both surfaces.

Distribution: south of the Yangtze River and southwest China.

Figure 4-14 *Phoebe sheareri* 紫楠

Phoebe zhennan **S. Lee et F. N. Wei** 楠木(桢楠) II (Figure 4-15)

Habit: trees, evergreen; trunk straight.

Bark: yellowish white; lenticels conspicuous.

Twig: slender, gray-yellowish or gray-brown villous or pubescent; buds densely gray-yellowish appressed villous.

Leaf: petiole hairy, elliptic, thinly leathery, abaxially pubescent and villous at veins, midrib abaxially conspicuously raised, veinlets abaxially slightly distinct or invisible.

Flower: inflorescence a cymose panicle, very patent, 7.5-12 cm long, lowest ramifications usually 2.5-4 cm long, slender, hairy.

Fruit: drupe, ellipsoid; persistent perianth lobes ovate, clasping fruit.

Distribution: southwest China.

* The species is similar to *Phoebe hui* W. C. Cheng ex Yen C. Yang 细叶楠 II closely, but the lateral veins of latter is obscure.

* The species is threatened by habitat loss, and so is under second-class national protection

Figure 4-15 *Phoebe zhennan* 楠木

in China. In the past, wood from this tree, referred to as 'nanmu' 楠木 in China was so valuable that only royal families could afford their use. Notably, whole logs of *P. zhennan* wood was used to create pillars for the Forbidden City.

***Phoebe chekiangensis* C. B. Shang** 浙江楠 Ⅱ (**Figure 4-16**)

  Habit: trees, evergreen.

  Bark: yellowish brown with distinct brown lenticels.

  Twig: angular, densely yellowish brown or gray blackish pubescent or tomentose.

  Leaf: obovate-elliptic or obovate-lanceolate, leathery, abaxially grayish brown pubescent and villous on veins, veinlets dense, abaxially conspicuous.

  Flower: inflorescence a panicle. Perianth lobes ovate, hairy on both surfaces.

  Fruit: drupe, ellipsoid-ovoid, persistent perianth lobes clasping base of fruit.

  Distribution: east China.

  * The species is similar to *P. sheareri* closely in the field, but leaf blade of the latter is much bigger and hair more conspicuous.

Figure 4-16 *Phoebe chekiangensis* 浙江楠

Phoebe bournei (Hemsl.) Yang 闽楠 Ⅱ (Figure 4-17)

Habit: tree, evergreen; trunk straight, few branched.

Bark: gray-white when old, yellowish brown when young.

Twig: hairy or glabrate.

Leaf: lanceolate or oblanceolate, leathery, abaxially pubescent and patent villous along veins, adaxially shiny, midrib abaxially raised, veinlets numerous, dense, conspicuously foveolate abaxially.

Flower: inflorescence a panicle, 3-7 cm long, lowest ramifications 2-2.5 cm long, hairy.

Fruit: drupe, ellipsoid or oblong; persistent perianth lobes clasping base of fruit.

Distribution: Fujian, Guangdong, Guangxi, Guizhou, Hainan, Hubei, and Jiangxi.

Figure 4-17 *Phoebe bournei* 闽楠

3. *Machilus* Nees 润楠属

Evergreen trees or shrubs. Leaves alternate, entire, pinnately veined. Inflorescence usually paniculate, terminal, subterminal, or arising from near base of branchlet. Flowers bisexual. Fruit fleshy, globose, subtended at base by persistent and reflexed perianth lobes.

***Machilus thunbergii* Sieb. et Zucc.** 红楠 (**Figure 4-18**)

Habit: trees, evergreen.

Bark: yellowish brown.

Twig: rough, but glabrous purple-brown when young. Buds scales golden brown or reddish brown, inner ones tomentose outside.

Leaf: petiole slender, reddish when fresh; leaf blade lustrous on both surfaces, obovate to obovate-lanceolate, leathery, glabrous on both surfaces, base cuneate, apex obtuse or abruptly cuspidate.

Flower: inflorescence arising from base of young shoot, 6-9 cm. Perianth lobes oblong.

Figure 4-18 *Machilus thunbergii* 红楠

Fruit: compressed globose, green becoming dark purple. Persistent perianth lobes reflexed.

Distribution: east China and south China.

Machilus leptophylla **Hand. −Mazz.** 薄叶润楠(华东楠) (**Figure 4-19**)

  Habit: tree, evergreen.

  Bark: grayish brown.

  Twig: thick, deeply brown, glabrous. Terminal buds subglobose, inner bud scales yellowish brown sericeous.

  Leaf: petiole glabrous; leaf blade abaxially glaucous, obovate-oblong, 12−22 cm long, thinly papery, veinlets sparse, inconspicuous, base cuneate, apex shortly acuminate.

  Flower: inflorescence panicles, many, congested on base of young branchlet. Flowers small, white when fresh.

  Fruit: drupe, globose.

  Distribution: east China, south China, Guizhou, and Hunan.

Figure 4-19 *Machilus leptophylla* 薄叶润楠

4. *Sassafras* J. Presl 檫木属

Deciduous trees. Leaves alternate, papery, pinninerved or triplinerved, dimorphic, unlobed or 2- or 3-lobed. Raceme terminal, pedunculate. Perianth lobes 6. Flowers unisexual or bisexual, pedicellate. Fertile stamens 9, all anthers 4-celled. Fruit dark blue, drupaceous, ovoid, glabrous, a shallow perianth cup at base.

Sassafras tzumu (Hemsl.) Hemsl. 檫木 (Figure 4-20)

Habit: trees, deciduous

Bark: yellow-green but gray-brown when mature, smooth, irregularly and longitudinally fissured.

Twig: reddish initially but blackish when dry, robust, glabrous. Terminal buds large, bud scales densely yellow sericeous outside.

Leaf: alternate, ovate or obovate, dimorphic, unlobed or 2- or 3-lobed; petiole always reddish when fresh, pinninerved or triplinerved, base cuneate, apex acuminate.

Flower: inflorescence a raceme terminal, appearing before leaves. Flowers small, yellow. Perianth tube short.

Figure 4-20 *Sassafras tzumu* 檫木

Fruit: subglobose, small, blue-black and white waxy when mature, seated on red shallow perianth cup.

Distribution: from Zhejiang to Sichuan, south of the Yangtze River.

5. *Litsea* Lam. 木姜子属

Inflorescence an umbel, or umbellate cyme or panicle; involucral bracts 4-6, decussate. Perianth tube segments usually 6 or abscent. Flowers unisexual, dioecious. Fertile stamens 9 or 12; anthers 4-celled, cells opening by lids. Fruit a drupe, berry-like, seated on perianth tube.

Litsea cubeba (Lour.) Pers. 山鸡椒(山苍子) (Figure 4-21)

Habit: shrubs or small trees, deciduous.

Bark: green or yellowish green.

Twig: slender, glabrous or sericeous-pubescent.

Leaf: alternate, lanceolate, oblong, or elliptic, papery; petiole glabrous; black when dry, but whitish abaxially, base cuneate, apex acuminate or acute.

Flower: inflorescence an umbel solitary or clustered, flowering before leaves or with leaves.

Fruit: subglobose, black at maturity.

Distribution: south provinces of the Yangtze River, Xizang and Taiwan.

Figure 4-21　*Litsea cubeba* 山鸡椒

* In the genus, *Litsea rotundifolia* var. *oblongifolia* (Nees) C. K. Allen 豺皮樟 is also common: bark cracked, yellow or slightly brown. Branchlets lenticels conspicuous. Petiole short; leaf blade ovate-oblong, base cuneate or obtuse, apex obtuse or shortly acuminate, inconspicuously reticulate-veined on both surfaces.

6. *Lindera* Thunb. 山胡椒属

Evergreen or deciduous. Leaves alternate, entire on margins or 3-lobed. Umbels singular and axillary. Flowers unisexual. Tepals 6; anthers 2-celled, introrse, dehiscing by flaplike. Fruit a drupe, perianth tubes inflated into a hypocarpium at base of fruit or cup-shaped.

Linderaglauca (Sieb. et Zucc.) Bl. 山胡椒(假死柴)(Figure 4-22)

Habit: shrubs or trees, deciduous.

Bark: smooth, gray or gray-white.

Twig: white-yellow, brown pubescent, later glabrate. Bud scales red on exserted parts.

Leaf: alternate, broadly elliptic, elliptic, obovate, papery, greenish and white pubescent

Figure 4-22　*Lindera glauca* 山胡椒

abaxially, green adaxially, pinninerved.

Flower: inflorescence an umbel axillary, 2-4 flowers.

Fruit: drupe, globose, black.

Distribution: south of the Yangtze River-Huaihe River, Henan, Gansu, Shandong, Shaanxi, Shanxi, and Taiwan.

* The species is similar to *Lindera angustifolia* W. C. Cheng 狭叶山胡椒, but the latter's young branchlets yellow-green, glabrous, leaf blade elliptic-lanceolate, apex rounded, bud scales keeled.

Lindera aggregate (Sims) Kosterm. 乌药 (Figure 4-23)

Habit: shrubs or small trees, evergreen.

Bark: grayish green.

Twig: blueish green, longitudinally striate, densely golden sericeous or laxly pubescent, gradually deciduous and glabrous, brown when dry. Terminal bud dense sericeous.

Leaf: alternate, pale abaxially, green and shiny adaxially, narrowly ovate to broadly elliptic, leathery, densely brown pubescent abaxially, trinerved, base rounded, apex acuminate.

Flower: inflorescence an umbel axillary, each with a bract and flowers 7. Tepals 6, yellow or yellowish green.

Fruit: ovate or sometimes subrounded, black.

Distribution: south of the Yangtze River, Hainan and Taiwan.

Figure 4-23 *Lindera aggregate* 乌药

4.4 Rosaceae 蔷薇科[①]

Leaves simple or compound, pinnate, alternate, stipulate. Flowers actinomorphic usually; sepals 5; petals 5, polypetalous; stamens indefinite; carpel 1 to many, apo- or syncarpous, situated on swollen receptacle. Fruit a follicle, pome, achene, or drupe, rarely a capsule.

There are 55 genera and 950 species in China.

* According to the fruit types, the family was traditionally divided into four subfamilies: Rosoideae 蔷薇亚科, Spiraeoideae 绣线菊亚科, Maloideae (Pomoideae) 苹果亚科, and Amygdaloideae (Prunoideae) 李亚科.

Ⅰ. Spiraeoideae 绣线菊亚科

Woody plants, rarely herbs. Leaves simple, rarely compound; exstipulate. Distinct or rarelyconnate carpels; ovules 2-many per carpel; overy superior, maturing as an aggregate of follicle or rarely as a capsule.

1. *Spiraea* Linn. 绣线菊属

Flowers in umbels, umbel-like racemes, corymbs, or panicles. Hypanthium campanulate or cupular. Sepals 5. Petals 5. Earpels usually 5, free. Fruit a follicle, bony, often dehiscent along adaxial suture.

***Spiraea prunifolia* Sieb. et Zucc. 李叶绣线菊**

Habit: shrubs, deciduous.

Twig: slender, reddish brown, turning gray-brown when old, pubescent but later glabrescent.

Leaf: petiole pubescent, ovate to oblong-lanceolate, pinnately veined, base cuneate, margin minutely sharply serrate from base or above middle to apex, apex acute.

Flower: inflorescence an umbel sessile. Flowers single in wild but double in cultivated. Hypanthium campanulate. Sepals triangular or ovate-triangular. Petals white, glabrous.

Fruit: follicle, glabrous.

Distribution: east China, north China, central China, Shaanxi, Sichuan, Taiwan, and Xizang.

* Sometimes, *Spiraea prunifolia* var. *simpliciflora* Nakai 单瓣李叶绣线菊(笑靥花)(Figure 4-24) also cultivated in garden usually, whose botanical features: leaf blade margin mostly minutely sharply serrate, flower single and small; follicle pubescent on adaxial suture.

[①] The rose family is arguably one of the six most economically important crop plant families, and includes fruits, such as apples, pears, peaches, apricots, plums, cherries, strawberries, blackberries, and roses among the crop plants belonging to the family. Many genera are also highly valued ornamental plants. These include trees, shrubs, herbaceous perennials and climbers.

Figure 4-24　*Spiraea prunifolia* var. *simpliciflora* 单瓣李叶绣线菊

Spiraea japonica Linn. f. 粉花绣线菊（Figure 4-25）

　　Habit：shrubs，deciduous.

　　Twig：slender，brownish to purple-brown，glabrous or pubescent when young.

　　Leaf：petiole pubescent，ovate or ovate-elliptic to lanceolate，abaxially usually pubescent on veins，base cuneate to rounded，margin crenate to doubly serrate，apex obtuse or acute.

　　Flower：inflorescence a corymb terminal on erect，densely. Flowers small. Hypanthium campanulate. Sepals erect in fruit. Petals usually pink，sometimes white. Disk annular.

　　Fruit：follicles divergent，glabrous or pilose on adaxial suture.

　　Distribution：south of the Yangtze River，Sichuan，Gansu，Xizang and Shaanxi.

2. *Exochorda* Lindl. 白鹃梅属

　　Shrubs deciduous. Stipules absent or small and caducous；leaf blade simple，margin entire or

Figure 4-25 *Spiraea japonica* 粉花绣线菊

serrate. Inflorescence a terminal raceme. Sepals 5, petals 5, imbricate, white, oblong to broadly obovate, base attenuate into a claw. Carpels 5, connate; ovary superior, 5 - loculed. Fruit a capsule.

Exochorda racemosa (Lindl.) Rehd. 白鹃梅 (Figure 4-26)

Habit: shrubs, deciduous.

Twig: red-brown but becoming brown later; buds dark purple, apex obtuse.

Leaf: petiole short or nearly absent, elliptic to oblong-obovate, glabrous on both surfaces, base cuneate, margin entire, rarely obtusely serrate above middle, apex obtuse or acute.

Flower: inflorescence a raceme, 6 - 10 - flowered. Hypanthium shallowly campanulate, glabrous. Petals white, obovate, base shortly clawed.

Fruit: capsule, glabrous.

Distribution: east China and Henan.

Figure 4-26 *Exochorda racemosa* 白鹃梅

II. Maloideae 苹果亚科

Woody plants with 2-5±connate carpels; ovules 1-2 per carpel; ovary inferior or semi-inferior, with adnate fleshy hypanthium, maturing as a pome or a pomelike berry.

3. *Pyracantha* M. Roem. 火棘属

Shrubs or small trees, evergreen, usually with thorny branches. Leaves simple, margin crenulate, serrulate, or entire. Inflorescence compound corymbs. Petals 5, spreading, white. Carpels 5, ovary 5-loculed, semi-inferior, with 2 ovules per locule. Fruit a pome, small.

Pyracantha fortuneana (Maxim.) H. L. Li 火棘 (Figure 4-27)

Habit: shrubs, evergreen.

Bark: grayish brown.

Twig: lateral branches short, thornlike; rusty pubescent when young, but dark brown and glabrescent later.

Leaf: obovate or obovate-oblong, both surfaces glabrous, base cuneate, margin serrate with teeth incurved, apex obtuse or emarginate.

Flower: inflorescence a compound corymb. Flowers small. Hypanthium campanulate. Petals white, apex rounded or obtuse.

Fruit: pome, orangish red or dark red, subglobose; sepals persistent, erect.

Distribution: from East China to southwest China, and other provinces.

Figure 4-27　*Pyracantha fortuneana* 火棘

4. *Crataegus* Linn. 山楂属

Shrubs or small trees, deciduous, armed usually. Leaves simple, margin serrate and lobed or partite, rarely entire. Hypanthium campanulate. Petals 5, white; carpels 1-5: Ovary inferior or semi-inferior, with 2 ovules per locule. Fruit a pome, with persistent sepals at apex.

***Crataegus pinnatifida* Bunge 山楂（Figure 4-28）**

Habit: trees, deciduous.

Bark: dark gray, longitudinally fissured.

Twig: purplish brown when young, but grayish brown later; buds purplish red, glabrous.

Leaf: broadly ovate or triangular-ovate, base truncate or broadly cuneate, with 3-5 pairs of lobes, margin sharply irregularly doubly serrate. Stipules falcate, margin serrate.

Flower: inflorescence a corymb, many flowered Hypanthium campanulate. Petals white, obovate or suborbicular. Ovary 5-loculed, with 2 ovules per locule.

Fruit: pome dark red, subglobose or pyriform, glabrous; sepals persistent.

Distribution: northeast China, north China, east China, Xinjiang and Shaanxi.

Figure 4-28 *Crataegus pinnatifida* 山楂

* *Crataegus pinnatifida* var. *major* N. E. Brown 山里红 is a long history cultivated plant as an important fruit tree, whose features are pedicel and peduncle pubescent, glabrate, fruit relatively large.

5. *Photinia* Lindl. 石楠属

Trees or shrubs, deciduous or evergreen. Leaves alternate, simple, margin serrate, rarely entire. Inflorescence terminal, umbellate or corymbose. Peduncle and pedicle with lenticles usually. Sepals 5, persistent, short. Petals 5, white. Fruit a pome.

Photinia serratifolia (Desfontaines) Kalkman 石楠 (Figure 4-29)

Habit: shrubs or trees, evergreen.

Bark: dark gray or grayish brown.

Twig: brown or reddish brown, but brownish gray later, glabrous; terminal buds red usually.

Leaf: petiole 2 – 4 cm long, glabrescent; leaf blade narrowly elliptic to obovate – elliptic, leathery, both surfaces glabrous when mature, margin conspicuously serrates, apex acuminate.

Flower: inflorescence a compound corymb terminal. Hypanthium cupular. Petals white, glabrous or villous.

Fruit: pome, red when immature, but brownish purple mature, globose.

Distribution: almost from the eastern part of China to the southwest and Gansu.

* Sometimes, *Photinia bodinieri* Lévl. 贵州石楠 is also popular in garden: thorns conspicuously in trunk and branchlets, peduncle and pedicle with hair, flowers larger, etc. In addition, *Photinia davidsoniae* Rehd. & Wils. 椤木石楠 has been merged into this species in FOC.

Figure 4-29 *Photinia serratifolia* 石楠

6. *Eriobotrya* Lindl. 枇杷属

Trees or shrubs, evergreen. Leaves simple; margin serrate or entire. Inflorescence terminal panicle, densely tomentose. Hypanthium cupular or obconical. Sepals 5. Petals 5, white or yellow. Ovary inferior; styles 2 - 5, connate at base and often pubescent. Fruit a pome with persistent incurved sepals, endocarp membranous, with 1 or 2 large seeds.

***Eriobotrya japonica* (Thunb.) Lindl. 枇杷 (Figure 4-30)**

Habit: trees, evergreen.

Bark: grayish yellow or gray.

Twig: yellowish brown, densely rusty or grayish rusty tomentose.

Leaf: lanceolate, oblanceolate, obovate, or elliptic - oblong, leathery, abaxially densely gray rusty tomentose, base cuneate, margin entire basally, remotely serrate apically, apex acute.

Flower: inflorescence a panicle, many flowered; peduncles, pedicels, sepals, with densely rusty tomentose. Petals white, fragrant.

Fruit: pome yellow or orangish yellow, rusty tomentose, soon glabrescent.

Distribution: native to Chongqing and Hubei, but cultivated in many areas of China.

Figure 4-30 *Eriobotrya japonica* 枇杷

7. *Sorbus* Linn. 花楸属

Trees or shrubs, usually deciduous. Leaves alternate, simple or pinnately compound. Inflorescences compound. Hypanthium campanulate. Sepals 5, petals 5. Fruit a pome, small.

Sorbus amabilis **Cheng ex T. T. Yü et K. C. Kuan** 黄山花楸（Figure 4-31）

Habit: trees, deciduous.

Bark: dark grey.

Twig: brown when young, blackish gray when old, terete, with small lenticels; buds narrowly ovoid, scales several, dark reddish brown.

Leaf: imparipinnate, stipules caducous, leaflets 5 - or 6 - paired, oblong or oblong - lanceolate, base obliquely rounded, margin coarsely sharply serrate, apex acuminate.

Flower: inflorescence terminal. Flowers small. Hypanthium campanulate. Petals white. Stamens 20, shorter than petals. Styles 3 or 4, equalling or lower than stamens, densely pubescent basally.

Fruit: pome, red, subglobose.

Distribution: east China mainly.

Figure 4-31 *Sorbus amabilis* 黄山花楸

Sorbus alnifolia (Sieb. et Zucc.) K. Koch 水榆花楸 (Figure 4-32)

Habit: trees, deciduous.

Bark: gray, smooth.

Twig: dark reddish brown, but dark grayish brown later, with white lenticels.

Leaf: simple, ovate to elliptic - ovate or suborbicular, lateral veins nearly parallel and terminating in marginal teeth, margin irregularly sharply doubly serrate or lobed.

Flower: inflorescence a compound corymb terminal, loosely 6 - 25 - flowered. Hypanthium campanulate. Petals white, base clawed. Filaments white, slightly shorter than petals. Styles 2, connate basally.

Fruit: pome, red, oblong, ovoid - oblong, or globose, without or with few minute lenticels, sepals caducous, leaving a small annular scar.

Distribution: northeast China, north China, east China, central China, Sichuan and Taiwan.

* The species is similar to *Sorbus caloneura* (Stapf) Rehd. 美脉花楸, but leaf blade of the latter elliptic, 8 - 12 cm long, lateral veins 10 - 18 pairs, pome globose with conspicuous lenticels.

Figure 4-32 *Sorbus alnifolia* 水榆花楸

8. *Pyrus* Linn. 梨属

Trees or shrubs, deciduous. Leaves alternate, simple, margin serrate or entire. Inflorescence a corymbose-racemose. Petals 5, clawed. Anthers red to purple-red. Ovary inferior, 2-5-loculed, with 2 ovules per locule; styles 2-5, free. Pome with juicy pulp, rich in stone cells.

***Pyrus calleryana* Dcne.** 豆梨 (**Figure 4-33**)

Habit: trees, deciduous.

Bark: grayish white or dark gray, ridged crack.

Twig: reddish brown, but grayish brown later, initially tomentose, soon glabrescent.

Leaf: broadly ovate or ovate, rarely narrowly elliptic, glabrous, base rounded or broadly cuneate, margin obtusely serrate, apex acuminate. Stipules caducous.

Flower: inflorescence a raceme umbel-like; peduncle and pedicel glabrous. Hypanthium cupular, glabrous. Sepals abaxially glabrous, adaxially tomentose. Petals white. Styles 2, glabrous.

Fruit: pome blackish brown with pale dots, globose; sepals caducous.

Figure 4-33 *Pyrus calleryana* 豆梨

Distribution: central China, east China and south China.

* *Pyrus calleryana* var. *integrifolia* T. T. Yü 全缘叶豆梨, margin entire. Native to Jiangsu and Zhejiang.

Pyrus pyrifolia (Burm. F.) Nakai 沙梨 (Figure 4-34)

Habit: trees, deciduous.

Bark: grayish white or dark gray, ridged crack.

Twig: purplish brown or dark brown; winter buds ovoid, scales tomentose.

Leaf: ovate-elliptic or ovate, base rounded or subcordate, margin spinulose-serrate, apex acute. Stipules caducous; petiole, peduncle, and pedicel tomentose, but glabrescent later.

Flower: inflorescence a raceme umbel-like, 6-9-flowered. Hypanthium cupular. Sepals 5. Petals white, ovate. Styles 5, rarely 4, glabrous.

Fruit: pome brownish, with pale dots, subglobose, sepals caducous.

Distribution: east China, south China, southwest China, Hunan and Hubei.

Figure 4-34 *Pyrus pyrifolia* 沙梨

9. *Malus* Mill. 苹果属

Trees or shrubs, deciduous or semievergreen. Inflorescences corymbose-racemose. Flowers pedicellate. Sepals 5, petals 5. Anthers yellow. Ovary inferior, 3-5-loculed; styles 3-5, connate at base. Pome with cartilaginous endocarp (core).

Malus hupehensis (Pamp.) Rehd. 湖北海棠 (Figure 4-35)

Habit: trees, deciduous.

Bark: grayish white.

Twig: initially dark green, but purple or purplish brown later.

Leaf: ovate or ovate-elliptic, sparsely puberulous when young, glabrescent, base broadly cuneate, rarely rounded, margin acutely serrulate, apex acuminate. Stipules caducous.

Flower: inflorescence a corymb. Hypanthium campanulate. Sepals triangular-ovate, as long as or slightly shorter than hypanthium. Petals pink, becoming white soon. Styles 3-5, tomentose basally.

Fruit: pome, yellowish green, tinged red, ellipsoid or subglobose.

Distribution: south of Yangtze River and Huaihe River, Gansu, Shaanxi, Shandong, and Shanxi.

Figure 4-35 *Malus hupehensis* 湖北海棠

Malus halliana Koehne 垂丝海棠 (Figure 4-36)

Habit: trees, deciduous.

Bark: grayish white.

Twig: purple or purplish brown, puberulous, but glabrescent later.

Leaf: ovate to elliptic, dark green and often tinged purple adaxially, base cuneate or subrounded, margin obtusely serrulate, apex long acuminate. Stipules caducous.

Flower: inflorescence a corymb. Pedicel pendulous, purple, slender, sparsely pubescent. Hypanthium glabrous abaxially. Petals pink. Styles 4 or 5, long tomentose basally.

Fruit: pome, purplish, pyriform or obovoid, sepals caducous.

Distribution: east China, southwest China, Hubei and Shaanxi.

Figure 4-36 *Malus halliana* 垂丝海棠

Malus pumila **Mill.** 苹果（Figure 4-37）

Habit: trees, deciduous.

Bark: dark gray.

Twig: purplish brown, robust, short, densely tomentose when young, but glabrous later.

Leaf: elliptic, ovate, or broadly elliptic, both surfaces densely puberulous when young, adaxially glabrescent, base broadly cuneate or rounded, margin obtusely serrate, apex acute.

Flower: inflorescence a corymb, 3-7-flowered. Hypanthium tomentose abaxially. Sepals both surfaces tomentose. Petals white, but pink in buds. Ovary 5-loculed, with 2 ovules per locule; styles 5, gray tomentose basally.

Fruit: pome, red or yellow, depressed-subglobose, sepals persistent.

Distribution: native to Europe, Central Asia, but commonly cultivated in temperate zone of China.

Figure 4-37 *Malus pumila* 苹果

10. *Chaenomeles* Lindl. 木瓜属

Shrubs to small trees, sometimes with thorny branches. Leaves simple, alternate, margin serrate or crenate. Flowers solitary or fascicled. Sepals 5, petals 5. Ovary 5-loculed; styles 2-5, connate at base. Fruit a pome, large, with many brown seeds.

***Chaenomeles sinensis* (Thouin) Koehne 木瓜 (Figure 4-38)**

Habit: shrubs or small trees, deciduous.

Bark: yellowish white, strips.

Twig: purplish red, unarmed, initially pubescent, soon glabrate, with pale lenticels.

Leaf: elliptic-ovate or elliptic-oblong, base broadly cuneate or rounded, margin aristate and sharply serrate, apex acute. Stipules small, ovate-oblong, herbaceous.

Flower: solitary, medium. Hypanthium campanulate. Sepals reflexed, fine serrate. Petals pinkish, obovate.

Fruit: pome, fragrant, dark yellow, narrowly ellipsoid, sepals caducous.

Distribution: from south of the Qinling Mountains - Huaihe River to south China, and Shandong.

Figure 4-38 *Chaenomeles sinensis* 木瓜

Chaenomeles speciosa (Sweet) Nakai 皱皮木瓜(贴梗海棠) (Figure 4-39)

Habit: shrubs, deciduous.

Twig: purplish brown or blackish brown, glabrous, with pale brown lenticels; thorn.

Leaf: ovate to elliptic, base cuneate to broadly cuneate, margin shortly serrate, apex acute or obtuse. Stipules reniform or suborbicular, large, herbaceous.

Flower: pedicel absent or short. Flowers precocious, fascicled on 2nd year branchlets. Hypanthium campanulate, glabrous. Sepals erect, apex obtuse. Petals scarlet, rarely pinkish or white.

Fruit: pome, fragrant, yellow or yellowish green; sepals caducous.

Distribution: the same as *C. sinensis*.

Figure 4-39 *Chaenomeles speciosa* 皱皮木瓜

III. Rosoideae 蔷薇亚科

Herbaceous or shrubby. Carpel 1, or 2 - many distinct carpels; ovules usually 1 - 2 per carpel; ovary superior, maturing as aggregates of achenes or drupelets, sometimes with accessory hypanthium or receptacle.

11. *Rosa* Linn. 蔷薇属

Shrubs, erect to climbing, mostly prickly. Leaves alternate, odd pinnate, rarely simple; stipules adnate or inserted at petiole, rarely absent. Flowers solitary or in a corymb. Sepals 5, petals 5, imbricate. Carpels free, numerous; styles free or connate at upper part. Fruit a hip (蔷薇果), formed from fleshy hypanthium. Achenes numerous, woody.

***Rosa multiflora* Thunb.** 野蔷薇(多花蔷薇)(**Figure 4-40**)

Habit: shrubs climbing, deciduous.

Twig: glabrous; prickles paired below leaves, curved, stout, flat.

Leaf: stipules pectinate, mostly adnate to petiole, shortly prickly; leaflets 5-9, obovate, oblong, or ovate, base rounded or cuneate, margin simply serrate, apex acute or rounded-obtuse.

Flower: numerous to form a corymb. Hypanthium subglobose, glabrous. Sepals 5, deciduous. Petals 5, semi-double or double, white, pinkish, or pink (in some cultivated plants), fragrant. Styles connate in column, glabrous.

Fruit: hip, red-brown or purple-brown, subglobose, glabrous, shiny.

Distribution: north China, east China, central China, south China and southwest China.

* The species is similar to *R. multiflora* var. *cathayensis* Rehd. et Wils. 粉团蔷薇, but the latter petals pink and single, flowers bigger.

Figure 4-40 *Rosa multiflora* 野蔷薇

Rosa chinensis Jacq. 月季花 (Figure 4-41)

Habit: shrubs, evergreen or semievergreen.

Twig: purple-brown, robust, subglabrous; prickles abundant to absent, curved, stout, flat.

Leaf: stipules mostly adnate to petiole; rachis and petiole sparsely prickly and glandular-pubescent; leaflets 3-5, rarely 7, broadly ovate or ovate-oblong, margin acutely serrate, apex long acuminate or acuminate.

Flower: fasciculate or solitary, slightly fragrant or not. Sepals 5, deciduous, margin entire or few pinnately lobed. Petals 5, double, red, pink, white, or purple. Styles free, pubescent.

Fruit: hip, red, ovoid or pyriform, glabrous.

Distribution: cultivated all over China.

* Daily, common people often confuse the species with *Rosa rugosa* Thunb. 玫瑰, but the latter leaflets 5-7(-9), rugose due to concave veins. *R. rugosa* is endangered as a wild plant by picking and uprooting.

Figure 4-41 *Rosa chinensis* 月季花

Rosa banksiae W. T. Ait. 木香花 (Figure 4-42)

Habit: shrubs, deciduous or semievergreen, climbing.

Twig: red-brown, with large and rigid prickles, glabrous; but cultivated plants sometimes not prickly.

Leaf: stipules caducous, free; leaflets 3-5, rarely 7, elliptic-ovate or oblong-lanceolate, leathery, margin depressed-serrulate, apex acute or slightly acute.

Flower: inflorescence an umbel or corymb. Pedicles slender, glabrous. Sepals 5, entire. Petals 5, double, white or yellow. Carpels numerous; styles free, densely pubescent.

Fruit: hip, orange or black-brown, globose or ovoid, glabrous, with deciduous sepals.

Distribution: Jiangsu, Henan, Hubei, Gansu and southwest China; also widely cultivated in China.

Figure 4-42 *Rosa banksiae* 木香花

12. *Kerria* Candolle 棣棠属

Shrubs deciduous. Leaves alternate; stipules caducous; leaf simple, margin doubly serrate. Sepals 5. Petals 5, yellow. Carpels 5-8, free; ovules 2. Fruit an achene, sepals persistent.

Kerria japonica (Linn.) Candolle 棣棠花 (Figure 4-43)

Habit: shrubs, deciduous.

Twig: green, usually arcuate, glabrous.

Leaf: triangular-ovate or ovate, base subcordate, rounded, or truncate, margin sharply doubly serrate, apex acuminate; Stipules caducous.

Flower: sepals persistent in fruit. Petals broadly elliptic, yellow, apex emarginate.

Fruit: achenes brownish black, obovoid or hemispheric, rugose.

Distribution: east China, southwest China, central China and other provinces.

* *Kerria japonica* f. *pleniflora* (Witte) Rehd. 重瓣棣棠花 is also common in garden, but petals double.

Figure 4-43 *Kerria japonica* 棣棠花

13. *Rubus* Linn. 悬钩子属

Shrubs. Stems usually with prickles or bristles. Flowers usually bisexual, incymose panicles, racemes, or corymbs, or several in clusters or solitary. Sepals 5, persistent. Petals 5. Carpels many, each carpel becoming a drupelet or drupaceous achene. Fruit a drupelet or drupaceous aggregate achene.

Rubus lambertianus Ser. 高粱泡 (Figure 4-44)

Habit: shrubs lianoid, semideciduous.

Twig: brown or reddish brown, with sparse, curved minute prickles.

Leaf: simple, with sparse, minute prickles; leaves blade broadly ovate, abaxially pilose, with sparse, minute prickles along midvein, margin distinctly 3-5-lobed or undulate, serrulate.

Flower: inflorescence a cymose panicle, terminal or axillary. Pedicel, calyx abaxially thinly pubescent; margin of inner sepals gray tomentose. Petals white.

Fruit: aggregate fruit red at maturity, subglobose, glabrous, with many drupelets.

Distribution: east China, central China, south China, southwest China, Henan and Taiwan.

Figure 4-44 *Rubus lambertianus* 高粱泡

Rubus parvifolius Linn. 茅莓 (Figure 4-45)

Habit: shrubs, deciduous.

Twig: grayish brown or reddish brown to blackish brown, with soft hairs and sparse, curved prickles.

Leaf: imparipinnate, 3-5-foliolate; petiole with soft hairs and minute prickles; blade of leaflets rhombic-orbicular or obovate, abaxially densely gray tomentose, margin unevenly coarsely serrate or coarsely incised-doubly serrate.

Flower: inflorescence a corymbose, terminal; rachis and pedicels pubescent, with minute prickles. Sepals erect, spreading. Petals pink to purplish red. Filaments white, linear.

Fruit: aggregate fruit red, ovoid-globose; pyrenes shallowly rugose.

Distribution: mostly part of China.

Figure 4-45 *Rubus parvifolius* 茅莓

Ⅳ. Prunoideae 李亚科

Woody plants, usually with 1 carpel or rarely 2-5 distinct carpels; ovules 1-2 per carpel; ovary superior, maturing as a drupe.

14. *Amygdalus* Linn. 桃属

Axillary buds in 3 (vegetative bud central, 2 flower buds to sides); flowers in early spring, sessile or nearly so, not on leafed shoots. Fruit a drupe, with a groove along one side; endocarp deeply grooved.

Amygdalus persica Linn. 桃 (Figure 4-46)

Habit: trees, deciduous.

Bark: dark reddish brown, scabrous and squamose with age, transversely lenticellate.

Twig: green but reddish on exposed side, slender, glabrous, with many small lenticels. Winter buds often 2 or 3 in a fascicle.

Leaf: oblong - lanceolate, elliptic - lanceolate, or obovate - oblanceolate, margin finely to coarsely serrate, apex acuminate.

Flower: solitary, occurring before leaves. Hypanthium shortly campanulate. Petals pink or white. Stamens many; anthers purplish red.

Fruit: drupe, color various, densely pubescent, ventral suture conspicuous; mesocarp color various, succulent, fragrant; endocarp large, surface longitudinally and transversely furrowed and pitted.

Distribution: native to north and center of China, but cultivated all over China.

Figure 4-46 *Amygdalus persica* 桃

15. *Armeniaca* Scop. 杏属

Axillary bud paralleling with flower bud; terminal winter bud absent. Leaves convolute when young. Fruit a drupe, hairy, with a conspicuous longitudinal groove, endocarp smooth.

***Armeniaca vulgaris* Lam.** 杏 (**Figure 4-47**)

  Habit: trees, deciduous.

  Bark: grayish brown, longitudinally splitting.

  Twig: brownish, glabrous; young branchlets reddish brown, with pale lenticels.

  Leaf: broadly ovate to orbicular-ovate, both surfaces glabrous, base broadly cuneate, rounded, or subcordate and with several nectaries, margin crenate, apex acute to shortly acuminate.

  Flower: solitary or occasionally paired, occurring before leaves. Hypanthium shortly cylindrical, purple-red. Sepals purplish green, reflexed after anthesis. Petals white, or pink. Ovary pubescent.

  Fruit: drupe, white, yellow, orange, often pubescent; endocarp nearly globose, with keel-like ribs on ventral side, surface scabrous or smooth.

  Distribution: native to west Asia, but cultivated all over China.

***Armeniaca mume* Sieb.** 梅 (**Figure 4-48**)

  Habit: trees, deciduous.

  Bark: grayish to tinged with green, corky ridges.

Figure 4-47 *Armeniaca vulgaris* 杏

Twig: young branchlets green, smooth, glabrous.

Leaf: ovate to obovate-oblanceolate, base broadly cuneate to rounded, margin usually acutely serrulate, apex caudate.

Flower: solitary or 2 in a fascicle, occurring before leaves, fragrant. Pedicel short or sessile. Hypanthium broadly campanulate. Sepals spreading after blooming. Petals white or pink.

Fruit: drupe yellow to greenish white, subglobose, pubescent; endocarp distinctly longitudinally furrowed on ventral and dorsal sides, surface pitted.

Distribution: native to southwest China, but cultivated all over China.

Figure 4-48 *Armeniaca mume* 梅①

---

① The plum blossom, which is known as the meihua (梅花), is one of the most beloved flowers in China and has been frequently depicted in Chinese art and poetry for centuries. The plum blossom is seen as a symbol of winter and a harbinger of spring. The blossoms are so beloved because they are viewed as blooming most vibrantly amidst the winter snow, exuding an ethereal elegance, while their fragrance is noticed to still subtly pervade the air at even the coldest times of the year. Therefore, the plum blossom came to symbolize perseverance and hope, as well as beauty, purity, and the transitoriness of life. In Confucianism, the plum blossom stands for the principles and values of virtue.

16. *Prunus* Linn. 李属

Axillary buds solitary. Flowers in early spring stalked, not on leafed shoots. Fruit glaucous, with a groove along one side, endocarp rough.

Prunus salicina Linn. 李 (Figure 4-49)

Habit: trees, deciduous.

Bark: dark gray, rough.

Twig: branchlets, petioles, pedicels, outside base of hypanthium glabrous or densely pubescent. Branches purplish brown to reddish brown; branchlets yellowish red.

Leaf: oblong-obovate, narrowly elliptic, margin doubly crenate and often mixed with simple gland-tipped teeth; Petiole apex with 2 nectaries.

Flower: usually 2-4 in a fascicle. Petals white, margin erose near apex. Ovary glabrous. Stigma disc-shaped.

Fruit: drupe, yellow or red, sometimes green or purple; exocarp waxy powder, endocarp rugose.

Figure 4-49 *Prunus salicina* 李

Distribution: cultivated all over China.

* *Prunus cerasifera* f. *atropurpurea* (Jacq.) Rehd. 紫叶李 is a common garden plant; leaf blade, pedicel, sepal and pistil all purple-red.

17. *Cerasus* Mill. 樱属

Leaves simple; petiole usually with 2 nectaries. Inflorescences axillary, fasciculate - corymbose or 1 - or 2 - flowered. Fruit a drupe, glabrous, not glaucous, without a longitudinal groove. Endocarp smooth or ± rugose.

Cerasus pseudocerasus (Lindl.) London 樱桃 (**Figure 4-50**)

Habit: small trees, deciduous.

Bark: grayish white to reddish brown, lenticels conspicuously.

Twig: young branchlets green, glabrous or pilose, but becoming grayish brown later.

Leaf: ovate, oblong-ovate, or long elliptic, abaxially pilose, margin acutely biserrate or incised serrate, teeth with a minute apical gland. Petiole apex with 1-3 large nectaries.

Flower: inflorescence a corymbose or subumbellate. Flowers occurring before leaves.

Figure 4-50 *Cerasus pseudocerasus* 樱桃

Hypanthium tubular, outside pilose. Sepals shorter than hypanthium. Petals white, apically emarginate or 2-lobed.

Fruit: drupe, red, subglobose; endocarp ± sculptured.

Distribution: from the Yellow River basin to the Yangtze River basin.

Cerasus serrulata (**Lindl.**) **London** 山樱花 (**Figure 4-51**)

Habit: trees, deciduous.

Bark: grayish brown to grayish black, lenticels conspicuously.

Twig: grayish white or tinged brown, glabrous.

Leaf: ovate-elliptic to obovate-elliptic, margin acuminately serrate or biserrate and teeth with a minute apical gland. Stipules linear. Petiole apex with 1-3 rounded nectaries.

Flower: inflorescence a corymbose-racemose or subumbellate. Pedicel glabrous. Hypanthium tubular, glabrous. Petals white or rarely pink, apex emarginate.

Fruit: drupe, purplish black, globose to ovoid.

Distribution: northeast China, east China, north China, and southwest China, and Shaanxi; cultivated all over China.

Figure 4-51 *Cerasus serrulate* 山樱花

* *Cerasus serrulate* var. *lannesiana* (Carr.) T. T. Yü et C. L. Li 日本晚樱 is widely cultivated in all over the world due to the brilliant ornamental features, and its leaf blade margin biserrate, petals pink, double.

Cerasus yedoensis (**Matsumura**) **A. V. Vassiljeva** 东京樱花(日本樱花) (**Figure 4-52**)

Habit: trees, deciduous.

Bark: silvery gray, lenticels conspicuously.

Twig: young branchlets green and pilose, turning pale purplish brown when old.

Leaf: elliptic to obovate, base rounded to rarely cuneate, margin sharply glandular serrate, apex acuminate to cuspidate.

Flower: inflorescence an umbellate-racemose, 3-6-flowered. Flowers occurring before leaves. Pedicel 2-3 cm, pubescent. Hypanthium tubular, outside pilose. Style base pilose.

Fruit: drupe, black, subglobose; endocarp slightly sculptured.

Distribution: native to Japan, but cultivated many metropolitan parks of China now.

Figure 4-52 *Cerasus yedoensis* 东京樱花

4.5　Calycanthaceae 蜡梅科

Shrubs, deciduous or evergreen. Branchlets fragrant; buds covered with scales or naked. Stipules absent. Leaves opposite, simple. Flowers bisexual, usually solitary, radially symmetric, usually fragrant. Tepals many, yellow, or yellowish white, spirally arranged on outer surface of a cup-shaped or urceolate receptacle. Fruit an achene, 1-seeded.

There are 2 genera and 7 species in China.

1. *Chimonanthus* Lindl. 蜡梅属

Vegetative buds with imbricate scales; flowers axillary, subsessile or very shortly pedicellate; tepals varying in size and shape from outer to inner but not distinctly dimorphic; fertile stamens 5-8.

Chimonanthus praecox (L.) Link 蜡梅 (Figure 4-53)

Habit: shrubs, deciduous.
Bark: brown.
Twig: grayish brown, quadrangular when young but becoming subterete, lenticellate.
Leaf: ovate, elliptic, broadly elliptic, papery to subleathery, abaxially glabrous except for

Figure 4-53　*Chimonanthus praecox* 蜡梅

occasional scattered trichomes on veins, adaxially roughly scabrous, base cuneate to rounded, apex acute.

Flower: solitary or paired, appearing generally before leaves, sweetly fragrant. Tepals 15-21, yellow but inner ones usually with purplish red pigment.

Fruit: achenes 3-11, brown, ellipsoid to reniform.

Distribution: from eastern Sichuan and Hubei to Zhejiang, cultivated throughout most of China.

2. *Calycanthus* Linn. 夏蜡梅属

Vegetative buds naked, hidden by base of petiole; flowers terminal, long pedicellate; tepals distinctly dimorphic in size and shape; fertile stamens 16-19.

Calycanthus chinensis **(W. C. Cheng et S. Y. Chang) W. C. Cheng et S. Y. Chang ex P. T. Li 夏蜡梅 (Figure 4-54)**

Habit: shrubs, deciduous.
Bark: glaucous or grayish brown, with convex lenticels.
Twig: glabrous or puberulous when young.
Leaf: broadly ovate-elliptic, both surfaces shiny, abaxially brown hispidulous but glabrescent,

Figure 4-54 *Calycanthus chinensis* 夏蜡梅

base broadly cuneate and slightly asymmetric, margin entire or irregularly serrulate, apex acute.

Flower: terminal, solitary. Tepals distinctly dimorphic; outer tepals white, flushed slightly pink, apex rounded; inner tepals pale yellow becoming white toward base, apex rounded and incurved.

Fruit: achene oblong, with silky trichomes.

Distribution: Zhejiang.

Taxonomically, Fabaceae 豆科 has been traditionally divided into three subfamilies, the Caesalpinioideae, Mimosoideae, and Papilionoideae (although sometimes these have been ranked as separate families, as in Caesalpiniaceae, Mimosaceae, and Papilionaceae). To follow the Hutchison's system, we adopted the viewpoint of three families in the textbook. The recognition of three families is based on characteristics particularly of the flower, including size, symmetry, aestivation of petals, sepals (united or free), stamen number and heteromorphy, pollen (single or polyads), but also presence of a pleurogram, embryo radicle shape, leaf complexity, and presence of root nodules. Differences in these characteristics led to the view that the Mimosaceae, Caesalpiniaceae and Papilionaceae are unique and distinct lineages in the Fables 豆目.

While there has been some disagreement as to whether Fabaceae should be treated as one family (composed of three subfamilies) or three, there is a growing body of evidence from morphology and molecules to support the legumes being one monophyletic family. This view has been reinforced not only by the degree of interrelatedness of taxonomic groups within the legumes compared to that between legumes and its relatives, but also by recent molecular phylogenetic studies showing strong support for a monophyletic family.

1a. Flowers actinomorphic, petals valvate in bud, free or united; anthers sometimes with a deciduous gland at apex ·· Mimosaceae 含羞草科
1b. Flowers ± zygomorphic, petals imbricate in bud
2a. Flowers slightly zygomorphic; corolla not papilionaceous, uppermost petal overlapped on each side by adjacent lateral petals (when these present); stamens with usually free filaments ·· Caesalpiniaceae 苏木科
2b. Flowers strongly zygomorphic (very rarely actinomorphic); corolla papilionaceous, standard outside wings, keel basally connate; stamens diadelphous (9+1) or monadelphous, rarely free ·· Papilionaceae 蝶形花科

4.6 Mimosaceae 含羞草科

Trees or shrubs; leaves bipinnate and stipulate, stipule may be modified into spines. Inflorescence cymose head or head. Flowers actinomorphic, hermaphrodite, small, tetra or pentamerous; calyx and corolla valvate; petals connate below, stamens number varies from 4 to many; carpel one. Fruit a legume.

There are 8 genera and 44 species native to China, but ca. 30 species were introduced.

Albizia Durazz. 合欢属

Leaves bipinnate; petiole and rachis with glands; leaflets small in numerous pairs or larger in few pairs, opposite. Inflorescence of globose head, or terminal panicle. Calyx campanulate or funnel-shaped, 5-toothed. Corolla funnel-shaped, upper part 5-lobed. Stamens numerous, connate into a tube at base. Legume broadly linear or oblong, straight, plano-compressed.

Albizia julibrissin Durazz. 合欢 (**Figure 4-55**)

Habit: trees, deciduous.

Bark: dark greenish grey and striped vertically when older.

Twig: angular; branchlet, leaf rachis, and inflorescence tomentose or pubescent.

Leaf: bipinnate, glands near base of petiole; 6-12 pairs of pinnae, each with 20-30 pairs of leaflets, obliquely linear to oblong, main vein close to upper margin, base truncate.

Flower: inflorescence a panicle terminal. flowers pink. Calyx tubiform, pubescent. Corolla lobes deltoid. Filaments pink.

Fruit: legume, strap-shaped, flat.

Figure 4-55 *Albizia julibrissin* 合欢

Distribution: south of the Yellow River basin and Taiwan.

Albizia kalkora (Roxb.) Prain 山槐(山合欢)(Figure 4-56)

Habit: trees, deciduous.

Bark: rough, greyish brown.

Twig: dark brown, pubescent, with conspicuous lenticels.

Leaf: bipinnate, pinnae 3 to 6 pairs, pinnules, 7 – 15 pairs, opposite, oblong to obovate, pubescent on both surfaces; stipules inconspicuous.

Flower: inflorescence a head, white, turning yellow, dimorphic; terminal flowers sessile, lateral flowers pedicellate.

Fruit: legume, linear to narrowly oblong, dark brown.

Distribution: east China, southwest China, central China, south China, Gansu, Henan.

Figure 4-56 *Albizia kalkora* 山槐

4.7 Caesalpiniaceae 苏木科

Leaves paripinnate; flowers zygomorphic; calyx and corolla 5, ascending imbricate; stamens 10 or less, free, gynoecium monocarpellary with marginal placentation. Fruit a legume.

There are 25 genera and ca. 110 species (including introduced) in China.

1. *Caesalpinia* Linn. 云实属

Trees, shrubs, or climbers, deciduous with prickles. Leaves bipinnate. Inflorescence raceme or panicle. Flowers bisexual, irregular. Sepals separate, imbricate, lowest one larger. Petals 5, often clawed, spreading, uppermost smaller. Stamens 10. Legume compressed or swollen to falcate.

Caesalpinia decapetala (Roth) Alston 云实 (Figure 4-57)

Habit: climbers, with copious prickles.
Bark: dull red.
Twig: branches, rachis of leaves, and inflorescence with recurved prickles and pubescent.
Leaf: bipinnate, pinnae 3-10 pairs, opposite, with prickles in pairs at base; stipules caducous; leaflets 8-12 pairs, oblong, membranous, both surfaces puberulent, glabrescent when old.

Figure 4-57 *Caesalpinia decapetala* 云实

Flower: inflorescence a raceme, terminal; rachis densely prickly. Sepals 5. Petals reflexed at anthesis, yellow, base shortly clawed. Filaments compressed at base, lanate in lower part.

Fruit: legume, chestnut-brown, shiny, fragile-leathery, glabrous, dehiscent and thickened to a narrow wing along ventral suture.

Distribution: south of the Yangtze River basin, Henan, Shaanxi and Taiwan.

2. *Gleditsia* Linn. 皂荚属

Trunk and branches usually with stout, simple or branched spines. Leaves paripinnate and/or bipinnate; leaflets numerous, base oblique or subsymmetrical, margin serrulate or crenate. Inflorescences axillary, rarely terminal, spikes or racemes, rarely panicles. Legume ovoid or elliptic, flat or subterete.

Gleditsia sinensis Lam. 皂荚（Figure 4-58）

Habit: trees, deciduous.

Bark: grayish white, or dark gray, lenticel conspicuous.

Twig: grayish to deep brown. Thorns robust, often branched.

Leaf: pinnate leaflets 6 – 14 pairs, ovate – lanceolate to oblong, papery, base rounded or cuneate, sometimes slightly oblique, margin serrate, apex acute.

Flower: inflorescence a raceme, axillary. Flowers polygamous, yellowish white. Sepals 4,

Figure 4-58 *Gleditsia sinensis* 皂荚

campanulate. Petals 4, white. Stamens 6-8. ovary hairy at base and on sutures.

Fruit: legume, brown or reddish brown, curved, strap-shaped, straight or twisted, with slightly thick pulp, swollen on both surfaces, often with farinose.

Distribution: south of the Yellow River basin, west to Sichuan, and south to Guangdong and Guangxi.

3. *Cercis* Linn. 紫荆属

Shrubs or trees, deciduous. Leaves simple, entire, veins palmate. Flowers bisexual, in solitary racemes or subumbellate clusters on branches of current year or older branches or trunks. Petals 5, zygomorphic, appearing like-papilionaceous. Stamens 10. Legume compressed, usually narrowly winged along ventral suture.

Cercis chinensis **Bunge** 紫荆（**Figure 4-59**）

Habit: shrubs or small trees, deciduous.

Bark: grayish white, lenticel conspicuous.

Twig: grayish white.

Leaf: suborbicular or triangular-orbicular, papery, both surfaces usually glabrous, base

Figure 4-59 *Cercis chinensis* 紫荆

shallowly to deeply cordate, margin membranous, apex acute.

Flower: purplish red, pink, or white, 2-10-clustered on old branches or especially on trunk; keel tinged with deep purple stripes.

Fruit: legume, compressed, sutures with narrowly wing, indehiscent.

Distribution: south of the Yellow River basin.

* The species is similar to *Cercis chingii* Chun 黄山紫荆, but the latter legume thick and hard, dehiscent, valves not winged, twisting upon dehiscence, with thick straight beak.

4.8　Papilionaceae 蝶形花科

Leaves usually imparipinnate or trifoliate. Flowers solitary or in raceme, very irregular (papilionaceous), posterior petal biggest and outermost. Petals 5, vexillary in aestivation. Stamens 10, usually diadelphous (9)+1 or (5)+(5), sometimes monadelphous, rarely 9. Fruit a legume.

There are 119 genera and 1100 species (including introduced) in China.

1. *Ormosia* Jacks. 红豆树属 Ⅱ

Trees or shrubs, evergreen. Leaves imparipinnate, paripinnate, or rarely simple. Inflorescence paniculate or racemose, axillary or terminal. Calyx campanulate, teeth 5. Corolla papilionaceous; petals clawed; standard suborbicular; wings and keel oblique, keel petals free. Stamens 10, free. Fruit a legume, suture without wing. Seeds red, shiny.

Ormosia hosiei Hemsl. et E. H. Wils. 红豆树 Ⅱ (Figure 4-60)

Habit: trees, evergreen.

Bark: grayish green, smooth.

Twig: green, yellowish brown pubescent, becoming glabrescent.

Figure 4-60　*Ormosia hosiei* 红豆树

Leaf: imparipinnate, leaflets 5-7, blades pale green abaxially, dark green adaxially, ovate or ovate-elliptic, apex acute or acuminate; both lateral veins and veinlets conspicuously reticulate when dried.

Flower: inflorescence a panicle. Corolla white or purplish; standard obovate, both wings and keel oblong. Style purple, filiform, curved; stigma oblique.

Fruit: legume, suborbicular, compressed, apex shortly beaked; valves subleathery. Seeds 1 or 2, red, shiny.

Distribution: south of the Qinling Mountains-Huaihe River.

Ormosia henryi Prain 花榈木 Ⅱ (Figure 4-61)

Habit: trees, evergreen.

Bark: grayish green, smooth, shallowly striate.

Twig: densely appressed tawny tomentose.

Leaf: imparipinnate, leaflets 5-7, blades elliptic or oblong-elliptic, abaxial surface and petiole densely appressed yellowish brown tomentose, base rounded, margin slightly repand, apex broadly rounded or acute.

Flower: panicle terminal, or raceme axillary, densely appressed brownish tomentose. Calyx campanulate. Corolla greenish white. Stamens free, unequal.

Figure 4-61 *Ormosia henryi* 花榈木

Fruit: legume, compressed, oblong, apex beaked; valve purplish brown, leathery, glabrous, internally septate.

Distribution: east China, south China and some parts of southwest China.

2. *Sophora* Linn. 槐属

Leaves imparipinnate; leaflets many, entire. Raceme terminal or axillary. Calyx campanulate, 5-lobed. Standard orbicular to oblanceolate; wings asymmetric or symmetric; keel similar to wings. Stamens 10, free or fused at base. Legume cylindric, moniliform.

Sophora japonica **L.** 槐树(国槐) (**Figure 4-62**)

Habit: trees, deciduous.

Bark: gray-brown, longitudinally striate.

Twig: green in current year, glabrous.

Leaf: imparipinnate, petiole inflated at base, bud hidden, leaflets 9–15, blades ovate-lanceolate or ovate-oblong, papery, glaucous and sparsely to densely pubescent abaxially.

Flower: inflorescence a panicle terminal. Corolla white or creamy yellow; standard broadly ovate, claw short; wings ovate-oblong; keel similar to wings. Stamens 10, unequal, free, persistent.

Figure 4-62 *Sophora japonica* 槐树

Fruit: Legume, green, moniliform, obviously constricted between seeds, indehiscent, fleshy.

Distribution: most part of China, commonly in north China.

* *Sophora japonica* 'pendula' 龙爪槐 is common in gardening, its branchlets curved, pendulous.

3. *Robinia* Linn. 刺槐属

Trees or shrubs, deciduous. Stipule bristlelike or spinelike. Leaves imparipinnate. Raceme axillary. Corolla white, pink, or purple; standard large, retroflexed; wings curved; keel incurved, blunt. Stamens diadelphous. Legume compressed, narrowly winged along ventral suture.

***Robinia pseudoacacia* Linn.** 刺槐(洋槐) (**Figure 4-63**)

Habit: trees, deciduous.

Bark: gray-brown to dark brown, longitudinally fissured, rarely smooth.

Twig: gray-brown, sparsely hairy, glabrescent; stipulate spines conspicuously.

Leaf: imparipinnate, leaflets 7-25, usually subopposite, leaflet blades oblong, elliptic, or ovate, base rounded to broadly cuneate, margin entire, apex rounded, retuse, and apiculate.

Flower: inflorescence a raceme axillary, pendulous, fragrant. Calyx obliquely campanulate. Corolla white, stipitate, papilionaceous; Stamens diadelphous, one opposite to standard free.

Figure 4-63 *Robinia pseudoacacia* 刺槐

Fruit: Legume, brown or with reddish brown stripes, linear-oblong, compressed, narrow wings along ventral suture; calyx persistent.

Distribution: Cultivated in all over China except Hainan and Xizang.

4. *Wisteria* Nutt. 紫藤属

Lianas, deciduous. Leaves imparipinnate. Raceme terminal, elongate, pendulous. Calyx broadly campanulate, 5-lobed. Corolla standard reflexed, with 2 basal calluses; wings free from keel. Stamens diadelphous; vexillary stamen distinct from other 9 or slightly connate at middle of sheath. Ovary stipitate. Legume linear to oblanceolate, leathery.

Wisteria sinensis (Sims) Sweet 紫藤 (**Figure 4-64**)

Habit: lianas, deciduous.

Bark: dark gray.

Twig: white villous when young, soon glabrescent.

Leaf: imparipinnate, 7-13-foliolate; leaflet blades elliptic-ovate to lanceolate-ovate, both surfaces appressed pubescent when young but glabrescent, base rounded to cuneate, apex attenuate

Figure 4-64 *Wisteria sinensis* 紫藤

to caudate.

Flower: inflorescence a raceme terminal or axillary. Flowers fragrant. Corolla purple or occasionally white; standard orbicular, glabrous, apex truncate. Ovary tomentose.

Fruit: legume, oblanceolate, tomentose, hanging persistently.

Distribution: cultivated widely and beyond its natural range.

5. *Indigofera* Linn. 木蓝属

Shrubs, shrublets, perennial herbs, deciduous. Trichomes typically medifixed (T-shaped), or rarely simple multicellular hairs. Leaves usually imparipinnate. Raceme axillary. Corolla usually reddish, sometimes white or yellow. Stamens 10, diadelphous, only vexillary 1 free; anthers uniform, basifixed or subbasifixed, hairy. Fruit a legume.

Indigofera fortunei Craib 华东木蓝 (Figure 4-65)

Habit: shrubs, deciduous.

Stem: grayish brown or gray.

Twig: striate, glabrous.

Leaf: imparipinnate, 7-15-foliolate; leaflet blades opposite, abaxially with sparse appressed medifixed trichomes but glabrescent later, adaxially glabrous.

Flower: inflorescence a raceme, bracts caducous. Calyx tubular with short hair. Corolla purple to pink, outside with dense appressed trichomes.

Fruit: legume, brown, cylindric, glabrous.

Distribution: east China, Hubei, Henan and Shaanxi.

Figure 4-65 *Indigofera fortunei* 华东木蓝

6. *Dalbergia* Linn. f. 黄檀属

Trees, or large woody climbers. Leaves imparipinnate, rarely simple; leaflets alternate, rarely opposite. Raceme or panicle. Flowers small, papilionaceous. Corolla white to pale green; petal clawed. Stamens 10, rarely 9, monadelphous. Legume compressed, linguiform. Seeds 1-4.

Dalbergia hupeana **Hance** 黄檀 (**Figure 4-66**)

Habit: trees, deciduous.

Bark: dull gray.

Twig: young shoots pale green, glabrous.

Leaf: imparipinnate; leaflets 9-11, elliptic to oblong-elliptic, base rounded, apex obtuse or slightly emarginate.

Flower: inflorescence a panicle terminal or extending into axils of uppermost leaves. Calyx campanulate, 5-toothed. Corolla white or light purple. Stamens 10, diadelphous (5+5).

Fruit: legume, oblong or broadly ligulate, reticulate opposite 1 or 2(or 3) seeds.

Distribution: east China, central China, south China and southwest China.

Figure 4-66 *Dalbergia hupeana* 黄檀

Dalbergia odorifera T. C. Chen 降香 Ⅱ (Figure 4-67)

Habit: trees, semievergreen.

Bark: brown or pale brown, rough, longitudinally splitting.

Twig: lenticels dense, small.

Leaf: imparipinnate; leaflets 9-13, ovate or elliptic, base rounded or broadly cuneate, apex obtuse to acuminate.

Flower: inflorescence a panicle axillary. Calyx campanulate. Corolla creamy white or pale yellowish. Stamens 9, monadelphous. Ovary narrowly elliptic.

Fruit: legume, ligulate-oblong.

Distribution: Hainan.

* In China, the precious wood from this species is known as huali or huanghuali. The wood was used to make higher-quality furniture during the late Ming and early Qing dynasties.

Figure 4-67 *Dalbergia odorifera* 降香

7. *Lespedeza* Michx. 胡枝子属

Shrubs, or perennial herbs. Leaves pinnately compound, 3-foliolate. Racemes axillary or flowers fasciculate; bracts persistent, 2; flowers often dimorphic, corollate or not. Legume ovoid, obovoid, or ellipsoidal, lenticular, indehiscent, reticulate veined, 1-seeded.

***Lespedeza buergeri* Miq.** 绿叶胡枝子

Habit: shrubs, deciduous.

Twig: sparsely hairy.

Leaf: 3-foliolate, leaflets ovate-elliptic, terminal one larger, abaxially adpressed hairy, adaxially glabrous, base slightly acute or obtuse-rounded, apex acute.

Flower: inflorescence a raceme axillary or in panicle at upper part of branchlet. Corolla pale yellowish green; standard nearly orbicular, shortly clawed; wings elliptic-oblong, sometimes apex slightly purple; keel with long clawed.

Fruit: legume, oblong-ovoid, villous, reticulate veined.

Distribution: east China, central China, southwest China and northwest China.

* *Lespedeza thunbergii* subsp. *formosa* (Vogel) H. Ohashi 美丽胡枝子 (Figure 4-68) is also common in field: leaflets adaxially puberulent or rarely glabrescent. Lateral calyx lobes nearly equal to or slightly shorter than calyx tube.

Figure 4-68 *Lespedeza thunbergii* subsp. *formosa* 美丽胡枝子

4.9 Styracaceae 安息香科

Trees or shrubs, stellate pubescent or scaly usually. Corolla lobes 4-8, basally connate, imbricate valvate; stamens twice more than corolla lobes.

There are 10 genera and 54 species in China.

1. *Sinojackia* Hu 秤锤树属 II

Pedicel elongated, slender, pendulous. Stamens 10-14, inserted near base of corolla; connective green, convex. Style elongate, subulate, obscurely 3-lobed.

Sinojackia xylocarpa **Hu 秤锤树 II（Figure 4-69）**

Habit: trees deciduous.
Bark: grayish brown.
Twig: grayish brown stellate pubescent. densely
Leaf: ovate in flower bearing branches, other obovate to elliptic; margin serrate, apex acute, penniveined 5-7 pairs.

Figure 4-69 *Sinojackia xylocarpa* 秤锤树

Flower: inflorescence 3 – 5 – flowered. Flowers white, pendulous, with stellate pubescent; pedicel ca. 2-4 cm; Calyx lobes triangular.

Fruit: ovoid, with a conical rostrum; woody, freckle on surface.

Distribution: endemic to Nanjing.

2. *Styrax* Linn. 安息香属

Ovary 3-locular when young, turning into 1-locular; seeds 1 or 2, without wing, rounded at both ends; pedicel not jointed. Fruit a drupe, indehiscent or 3-valvate dehiscent.

Styrax dasyanthus Perk. 垂珠花（**Figure 4-70**）

Habit: trees deciduous.

Bark: grayish brown or dark.

Twig: densely gray-yellow stellate pubescent, but purple and glabrous later.

Leaf: alternate, obovate – elliptic to elliptic, leathery, base cuneate, margin denticulate and slightly revolute, apex acute to shortly acuminate.

Flower: inflorescence a paniculate, terminal or axillary. Corolla tube, lobes oblong to oblong-lanceolate. Filaments expanded, free parts basally densely white villose.

Figure 4-70 *Styrax dasyanthus* 垂珠花

Fruit: drupe, ovoid, densely grayish stellate tomentose.

Distribution: east China, central China and southwest China.

4.10 Symplocaceae 山矾科

Shrubs or trees, evergreen, or rarely deciduous. Stamens 12 to many, adnate to base of corolla tube. Ovary inferior or semi-inferior, 2-5-locular. Fruit a drupe.

There are 1 genus and 42 species in China.

Symplocos Jacq. 山矾属

The morphological characteristics are the same as the family's.

***Symplocos sumuntia* Buchanan-Hamilton ex D. Don 山矾 (Figure 4-71)**

Habit: trees, evergreen.

Bark: grayish brown.

Twig: brown, usually glabrous, lenticellate.

Leaf: elliptic to narrowly ovate, thinly leathery, apex caudate, pinninerved 4-8 pairs, becoming yellowish green when dry.

Figure 4-71 *Symplocos sumuntia* 山矾

Flower: raceme with pubescent. Calyx lobes triangular-ovate, glabrous. Corolla white, 5-lobes with slightly hairs. Stamens connate basally. Disc glabrous, annular, glabrous.

Fruit: drupe, persistent erect calyx lobes.

Distribution: south of the Yangtze River—Huaihe River and Taiwan.

Symplocos paniculata (**Thunb.**) **Miq.** 白檀 (**Figure 4-72**)

Habit: shrubs or small trees, deciduous.

Bark: gray, or whitish gray.

Twig: sparselypilose when young, but old glabrous.

Leaf: ovate or obovate, membranous to thinly papery, margin sharply glandular dentate, apex acuminate to acute, pinninerved 4-10 pairs.

Flower: panicles terminal. Corolla white, fragrant. Calyx lobes light yellow. Stamens many, connate basally. Ovary obconic, 2-locular.

Fruit: drupe, bluish with persistent erect calyx lobes, ovate-globose.

Distribution: northeast China, north China, Taiwan and south of the Yangtze River.

Figure 4-72 *Symplocos paniculata* 白檀

4.11　Cornaceae 山茱萸科

Leaves opposite or alternate. Flowers bisexual or unisexual. Ovules pendulous, 1 per locule. Fruit a drupe or rarely berry, or syncarpous compound fruit.

There are 1 genus and 25 species in China.

Cornus Linn. 山茱萸属

The morphological characteristics are the same as the family's.

Cornus hongkongensis Hemsley 香港四照花（Figure 4-73）

Habit: trees or shrub, evergreen.

Bark: grayish brown, smooth.

Twig: green or purplish green, sparsely pubescent or glabrous, lenticel obscure.

Leaf: opposite, elliptic, oblong-elliptic, or obovate-oblong; papery; base cuneate, apex shortly acuminate, margin entire.

Figure 4-73　*Cornus hongkongensis* 香港四照花

Flower: capitate cyme globose, with 4 yellowish or white bracts, petal-like. Petals 4, yellow; calyx tube, 4-lobes.

Fruit: compound fruit red at maturity, globose.

Distribution: from south China to southwest China.

Cornus officinalis Sieb. et Zucc. 山茱萸 (Figure 4-74)

Habit: trees or shrubs, deciduous.

Bark: grayish brown.

Twig: green when young, but turning dark brown later.

Leaf: opposite, ovate-lanceolate or ovate-elliptic, adaxially sparsely pubescent but abaxially dense light brown soft trichomes.

Flower: umbellate inflorescence axillary, bracts 4. Flowers yellow; calyx 4, triangular; petals 4, ovate.

Fruit: drupe, red when mature, narrowly ellipsoid.

Distribution: east China, central China, Gansu and Shaanxi.

Figure 4-74 *Cornus officinalis* 山茱萸

Cornus controversa Hemsley 灯台树 (Figure 4-75)

Habit: trees, deciduous.

Bark: dark gray or yellowish gray, smooth.

Twig: purplish, glabrous.

Leaf: alternate, broadly ovate to broadly elliptic-ovate, apex acute or acuminate, base rounded, grayish green abaxially with sparsely pubescent, pinninerved 6 or 9 pairs.

Flower: corymbose cyme terminal. Flowers bisexual, white; calyx teeth, triangular; petals 4, long-lanceolate.

Fruit: drupe, purplish red to bluish black.

Distribution: many areas of China.

Figure 4-75　*Cornus controversa* 灯台树

Cornus walteri Wangerin 毛梾 (Figure 4-76)

Habit: trees, deciduous.

Bark: dark gray, rectangularly splitting.

Twig: green when young, but turning greenish white to grayish black, glabrous.

Leaf: opposite, narrowly elliptic to broadly ovate, apex acuminate, base cuneate, abaxially with grayish white trichomes, pinninerved 4-5 pairs.

Flower: corymbose cymes terminal. Flowers fragrant, white. Calyx lobes triangular. Petals oblong-lanceolate. Stamens 4. Ovary with short pubescent.

Fruit: drupe, globose, black.

Distribution: most parts of China.

Figure 4-76 *Cornus walteri* 毛梾

4.12 Alangiaceae 八角枫科

Trees or shrubs, deciduous, sometimes spiny. Leaves alternate, simple, estipulate, bases often oblique. Petals as many as calyx lobes, linear to lorate. Ovary inferior. Fruit a drupe, crowned with persistent calyx and disk.

There are 1 genus and 11 species in China.

Alangium Lam. 八角枫属

The morphological characteristics are the same as the family's.

Alangium chinense (Lour.) Harms 八角枫 (Figure 4-77)

Habit: shrubs or small trees, deciduous.

Bark: grayish white.

Twig: flexuous, pubescent and purple when young, but glabrescent later.

Leaf: ovate or orbicular to cordate, strongly 3-5-veined at base, base oblique usually, margin entire or with few shallow lobes, apex acuminate.

Flower: inflorescence an axillary cyme. Calyx lobes 4-7, shortly dentate. Petals valvate, lanceolate. Stamens as many as petals. Filaments and style with short hair.

Fruit: drupe, ovoid, small.

Figure 4-77 *Alangium chinense* 八角枫

Distribution: south of the Qinling Mountains-Huaihe River, Taiwan and Xizang.

4.13 Nyssaceae 蓝果树科

Trees dioecious; winter buds with scales; Leaves alternate, simple, estipulate. Flowers in terminal or axillary; inflorescence heads, racemes, or umbels. Calyx tube adnate to ovary in bisexual or female flowers. Fruit a drupe or samara.

There are 3 genera and 10 species in China.

1. *Camptotheca* Dec. 喜树属

Leaves opposite, pinnate vein. Inflorescence heads terminal or axillary. Bracts 2 in each flower; flower disc apparently. Fruit a samaralike.

Camptotheca acuminata **Dec.** 喜树 Ⅱ (Figure 4-78)

Habit: trees, deciduous.

Bark: gray, deeply furrowed.

Twig: purplish, villous when young, but glabrous later.

Leaf: alternate, oblong-ovate to oblong-elliptic, papery, pinnate vein 10-11 pairs, base subrounded, margin entire, apex acute, petiole red with pubescent.

Flower: inflorescence a head, terminal or axillary. Flowers small, monecious. Calyx cup-shaped. Petals 5, light green. Disk conspicuous. Stamens 10, 2 whorls.

Fruit: gray-brown, thinly winged.

Distribution: east China, central China, south China and southwest China.

Figure 4-78 *Camptotheca acuminata* 喜树

2. *Davidia* Baill. 珙桐属

Trees, deciduous. Leaves alternate. Inflorescence a head, terminal, globose, pedunculate; bracts 2, white, larger one pendulous. Ovary 6-10-loculed. Fruit a drupe, usually solitary.

Davidia involucrata Baill. 珙桐 I （Figure 4-79）

Habit：trees, deciduous.

Bark：dark gray or dark brown, often split into irregular flakes.

Twig：purplish green in current year, glabrous, turning dark brown or dark gray.

Leaf：alternate, leaf blade adaxially bright green, broadly ovate, strongly veined, base cordate, margin dentate-serrate with acuminate teeth, apex acuminate.

Flower：inflorescence a head, terminal. Bracts 2, opposite, ovate to oblong-obovate. Male flower without calyx and petals; Female or bisexual flower with inferior ovary.

Fruit：drupe solitary, pear-shaped.

Distribution：southwest China.

* *Davidia involucrata* var. *vilmoriniana* (Dode) Wangerin 光叶珙桐 is similar to the species, but the former leaves abaxially glabrous or scarcely pubescent when young, sometimes abaxially glaucous.

Figure 4-79 *Davidia involucrata* 珙桐

4.14 Araliaceae 五加科

Inflorescence usually umbel or head. Ovary inferior, ovules 2 per locule, 1 mature. Fruit a drupe or berry.

There are 23 genera and 180 species in China.

1. *Kalopanax* Miq. 刺楸属

Trees, deciduous. Stem and branch often armed with prickles. Leaves simple, palmately lobed. Ovary 2-carpellate; styles united at base, 2-cleft apically. Fruit a drupe.

***Kalopanax septemlobus* (Thunb.) Koidz. 刺楸 (Figure 4-80)**

Habit: trees, deciduous; trunk with stout prickles.

Bark: dark grayish brown.

Twig: numerous prickles.

Figure 4-80 *Kalopanax septemlobus* 刺楸

Leaf: alternate, Leaf blade suborbicular, papery, palmated 5-7-lobed, base cordate or rounded to nearly truncate, margin serrate, apex acuminate.

Flower: inflorescence an umbel. Flowers small; corolla white or yellowish green.

Fruit: dark, blue when maturity.

Distribution: from northeast China to south China.

2. *Eleutherococcus* Maxim. 五加属

Usually prickly. Leaves palmately compound or trifoliolate. Panicle of umbel or a solitary umbel. Petals 5, valvate. Fruit a berry-like.

Eleutherococcus trifoliatus (Linn.) S. Y. Hu 白簕 (Figure 4-81)

Habit: shrubs, scandent or climbers.

Twig: recurved prickles, scattered.

Leaf: palmated, leaflets 3-5, ovate to elliptic-ovate, papery, base cuneate, margin serrulate, apex acute or acuminate. Petiole with prickles.

Flower: inflorescence a terminal raceme of umbel or a compound umbel. Calyx 5 teeth. Petals 5. Stamens 5. Ovary 2-carpellate, styles united to middle.

Figure 4-81 *Eleutherococcus trifoliatus* 白簕

Fruit: globose, laterally compressed, dark black.

Distribution: central China, south China, southwest China and Taiwun.

3. *Fatsia* Dec. et Planch. 八角金盘属

Shrubs or small trees, evergreen, unarmed. Leaves simple, palmately 7-9-lobed. Styles 5 or 10, free. Fruit a subglobose drupe.

***Fatsia japonica* (Thunb.) Dec. et Planch. 八角金盘 (Figure 4-82)**

Habit: shrubs, evergreen.

Twig: young branches, leaves, and inflorescences densely woolly tomentose, glabrescent later.

Leaf: orbicular, leathery, with 7-9 deeply cleft, both surfaces glabrous, base cordate, margin crenate to crenate-serrate, teeth blunt, apex acuminate.

Flower: inflorescence a panicle of umbel with numerous white flowers. Ovary 5-carpellate; styles 5, free.

Fruit: drupe, globose, dark black when mature.

Distribution: widely cultivated in east China.

Figure 4-82 *Fatsia japonica* 八角金盘

4.15 Caprifoliaceae 忍冬科

Shrubs deciduous usually. Leaves opposite, stipules absent usually. Corolla epigynous, gamopetalous; lobes 4 or 5, spreading, sometimes bilabiate, aestivation imbricate. Ovary inferior. Fruit a berry, berry-like or capsule.

There are 5 genera and 66 species in China.

1. *Viburnum* Linn. 荚蒾属

Evergreen or deciduous; winter buds naked or enveloped scales. Leaves opposite. Corolla white, campanulate, hypocrateriform, or tubular. Styles short.

***Viburnum odoratissimum* Ker Gawler** 珊瑚树（**Figure 4-83**）

Habit: shrubs or small trees, evergreen.

Bark: gray-brownish with raised lenticels.

Twig: current year green or reddish, glabrous; but gray or gray-brownish, terete, glabrous, raised lenticels in 2nd year.

Figure 4-83 *Viburnum odoratissimum* 珊瑚树

Leaf: opposite, stipules absent. Leaf blade intense green, elliptic to oblong, both surfaces glabrous, pinnate vein, base broadly cuneate, margin irregularly serrate except at base or subentire, apex shortly acute.

Flower: inflorescence a paniculate, pyramidal, terminal. Flowers fragrant. Calyx green, glabrous. Corolla white, later yellow-whitish. Stamens inserted at apex of corolla tube.

Fruit: berrylike drupe, ovoid or ovoid-ellipsoid, initially turning red, maturing nigrescent.

Distribution: south China.

* *Viburnum odoratissimum* var. *awabuki* (K. Koch) Zabel ex Rümpler 日本珊瑚树 is similar to the species, but the former petiole reddish, leaf blade lustrous, elliptic-obovate, thickly leathery; inflorescence axes glabrous; corolla campanulate. This variety is commonly cultivated in east China.

Viburnum dilatatum Thunb. 荚蒾 (Figure 4-84)

Habit: Shrubs, deciduous.

Twig: grayish brown, densely stellate-pubescent.

Leaf: opposite, stipule absent, leaf blade broadly obovate to ovate, papery, abaxially yellowish pubescent and stellate-pubescent, adaxially adpressed hairy, pinnate, 6-8 veins, margin serrate, apex acute.

Flower: inflorescence a compound umbel-like cyme. Corolla white, stellate-pubescent.

Figure 4-84 *Viburnum dilatatum* 荚蒾

Fruit: berrylike drupe, red, ellipsoid-ovoid.

Distribution: south of the Yangtze river Basin, Shaanxi and Taiwan.

2. *Lonicera* Linn. 忍冬属

Shrubs or climbers, deciduous or evergreen. Leaves opposite. Calyx 5 – lobed, base occasionally with a collarlike emergence. Corolla campanulate or funnelform, 5 – lobed, or bilabiate; often shallowly to deeply gibbous on ventral side toward base. Fruit a berry.

***Lonicera japonica* Thunb.** 忍冬(金银花)(**Figure 4-85**)

Habit: climbers, semievergreen.

Twig: hollow; spreading stiff hairs and glandular hairs in branche, petiole and peduncle.

Leaf: opposite, leaf blade ovate or oblong to lanceolate with densely hairy both sides, base rounded, margin ciliate, apex acute.

Flower: fragrant, paired and axillary toward apices of branchlets. Calyx lobes triangular. Corolla bilabiate, white, becoming yellow. Stamens and style glabrous.

Fruit: berry, black when mature, globose.

Distribution: many areas of China.

Figure 4-85 *Lonicera japonica* 忍冬

Lonicera maackii (Rupr.) Maxim. 金银忍冬(金银木) (Figure 4-86)

Habit: shrub, deciduous.

Twig: brown, hollow pith, pubescent with interspersed minute glands.

Leaf: ovate-lanceolate usually, abaxially sparsely strigose, adaxially sparsely pubescent or subglabrous, base broadly cuneate to rounded, margin ciliate, apex acute.

Flower: fragrant, axillary paired flowers, bracts linear. Calyx campanulate, lobes broadly triangular. Corolla bilabiate, purplish, or white at first, later yellow, upper lip 4-lobed, lower lip recurved. Stamens and style exserted from corolla tube.

Fruit: berry, dark red.

Distribution: north China, east China, central China and northeast China.

Figure 4-86　*Lonicera maackii* 金银忍冬

4.16 Hamamelidaceae 金缕梅科

Indumentum usually of stellate hairs or stellate or peltate scales. Stipules paired usually. Ovary inferior, 2-locular, carpels free at apex. Fruit a capsule with persistent styles.

There are 18 genera and 74 species in China.

1. *Liquidambar* Linn. 枫香树属

Trees, deciduous. Leaves alternate, palmately 3-7-lobed. Inflorescence usually a globose head. Capsule woody, dehiscing loculicidally by 2 valves; pericarp thin; styles persistent.

***Liquidambar formosana* Hance** 枫香树 (Figure 4-87)

Habit: trees, deciduous.

Bark: gray-brown, irregular cracked.

Twig: pubescent or glabrous; buds ovoid with puberulent.

Leaf: broadly ovate, palmately 3-lobed and 3-veined, base rounded, margin glandular serrate, apex caudate-acuminate. Stipules red, linear, caducous.

Figure 4-87 *Liquidambar formosana* 枫香树

Flower: male inflorescence a short spike. Female inflorescence a head.

Fruit: infructescence globose. Capsule with persistent styles, spike-like.

Distribution: south of the Yellow River, Hainan and Taiwan.

2. *Loropetalum* R. Brown 檵木属

Rustystellate hairs. Sepals usually 4. Petals 4, straplike. Stamens 4, anther thecae 2-sporangiate, each dehiscing by 2 valves. Capsule dehiscing by two 2-lobed valves

Loropetalum chinense (R. Brown) Oliver 檵木 (Figure 4-88)

Habit: shrubs or small trees, evergreen.

Twig: stellately pubescent.

Leaf: ovate or elliptic, discolorous, abaxially densely stellately pubescent, adaxially sparsely pubescent, base asymmetrical, margin entire, apex acute, veins 4-8 pairs.

Flower: inflorescence a short raceme. Peduncle stellately pubescent; bracts linear. Floral cup cupular, stellately pubescent. Petals 4, white, linear. Stamens 4. Ovary inferior.

Fruit: capsule ovoid, stellately tomentose, hairs brown, adnate to floral cup.

Distribution: east China, south China and southwest China.

* *Loropetalum chinense* f. *rubrum* H. T. Chang 红花檵木 (Figure 4-88 bottom right) is

Figure 4-88 *Loropetalum chinense* 檵木

similar to the original species, but its petals are usually purple-red or red. It is widely cultivated in south cities of China.

3. *Parrotia* C. A. Mey. 银缕梅属

Small trees, deciduous. Leaves pinnately veined, undivided. Petal absent, stamens variable in number. Ovary semi-inferior ovule 1 per locule. Fruit a capsule.

Parrotia subaequalis (H. T. Chang) R. M. Hao et H. T. Wei 银缕梅 I (Figure 4-89)

Habit: small trees, deciduous.
Bark: gray, irregular cracked.
Twig: stellately pubescent, glabrescent later.
Leaf: stipules narrowly lanceolate, petiole stellately pubescent, leaf blade broadly obovate or elliptic with stellately pubescent, margin sparsely sinuate dentate, apex obtuse, pinnate, veins 4 or 5 pairs.
Flower: inflorescences a capitate spike. Floral bracts ovate with densely stellately pubescent. Bisexual flowers 4 or 5, margin irregularly toothed, persistent. Stamens erect in buds. Ovary semi-inferior.
Fruit: capsule, subglobose, dehiscing by 2 valves, floral cup persistent; styles persistent.
Distribution: Anhui, Jiangsu and Zhejiang.

Figure 4-89 *Parrotia subaequalis* 银缕梅

4.17 Platanaceae 悬铃木科

Trees deciduous. Leaves alternate, palmately lobed and subpalmately veined. Inflorescence a globose-capitate. Infructescence a capitate or globose coenocarpium composed of numerous achenes.

There are 1 genus and 3 species in China (all introduced).

Platanus Linn. 悬铃木属

The morphological characteristics are the same as the family's.

***Platanus acerifolia* (Ait.) Willd.** 二球悬铃木 (Figure 4-90)

Habit: trees, deciduous.

Bark: pale brown, gray, or white, smooth, exfoliating in plates.

Twig: densely gray-yellow tomentose when young, old ones red-brown, glabrous.

Figure 4-90 *Platanus acerifolia* 二球悬铃木

Leaf: broadly ovate, 3-7-lobed, gray-yellow pubescent on both surfaces when young, principal veins 3 or 5, base subcordate or truncate, lobes entire or coarsely 1- or 2-dentate at margin.

Flower: flowers 4-merous usually. stamens longer than petals; anther connective peltate, pubescent.

Fruit: fruiting branchlet with 2 infructescences usually. Infructescence a capitate. Achene with persistent style spiniform.

Distribution: widely cultivated elsewhere.

* In *Platanus*, *Platanus occidentalis* Linn. 一球悬铃木 and *Platanus orientalis* Linn. 三球悬铃木 also cultivated in many areas of China, the former and the later with 1 and 3 infructescences in the fruiting branchlet, respectively.

4.18　Buxaceae 黄杨科

Shrubs or small trees, evergreen. Leaves simple, exstipulate. Flowers small, regular, unisexual. Petal absent usually. Ovules 2 per locule, pendent, anatropous, bitegmic, crassinucellar. Fruit a loculicidal dry capsule or a fleshy berry.

There are 3 genera and 28 species in China.

Buxus Linn. 黄杨属

Leaves opposite, margin entire, venation pinnate; female flower solitary, apical on inflorescence. Fruit a dry capsule, loculicidal, splitting into 3 valves; style persistent.

Buxus sinica (Rehd. et Wils.) Cheng 黄杨 (Figure 4-91)

Habit: shrubs or small trees, evergreen.

Bark: grayish white, longitudinally ribbed regularly.

Twig: young branch tetragonous, but becoming terete when old with longitudinally ribbed.

Leaf: varied in shape and size, broadly elliptic, broadly obovate, obovate, or elliptic-lanceolate to lanceolate, shining adaxially, glabrous on both surfaces or puberulent along basal half of midrib.

Flower: inflorescence a capitate axillary. Petal absent; sepals 6, 2 whorls. Female flower: style 3, thick and compressed.

Fruit: capsule, subglobose, black when mature; style persistent.

Distribution: north China, east China, southwest China, Gansu and Shaanxi; but cultivated widely in garden.

* the species is similar to *Buxus bodinieri* Lévl. 雀舌黄杨, but the latter is spatulate or obovate, midrib and lateral veins distinctly prominent; branchlets slender, glabrous.

Figure 4-91 *Buxus sinica* 黄杨

4.19   Salicaceae 杨柳科

Trees or shrubs, dioecious, rarely polygamous. Leaves alternate, simple; stipule persistent or caducous. Catkin erect or pendulous; each flower usually with a cupular disc or 1 or 2 nectariferous glands; usually lacking a normal perianth. Fruit a capsule; seeds with long hairs.

There are 3 genera and 347 species in China.

1. *Populus* Linn. 杨属

Growth monopodial, bud with several outer scales, terminal bud present; both male and female catkins pendulous; disc cupular; leaf blade usually 1-2 × as long as wide.

***Populus ×canadensis* Moench** 加杨（**Figure 4-92**）

Habit: trees, deciduous; crown ovoid.

Bark: whitish first smooth, later barky black with deep furrows.

Twig: yellowish to light gray with leaf scars; buds heaped and very thick, greenish and pointed.

Leaf: deltoid - ovate, abaxially greenish, adaxially dull green, base truncate or broadly cuneate, with 1 or 2 glands or not, margin crenate, apex acuminate.

Flower: inflorescence a catkin. Male flower: disc yellowish green, margin entire, stamens reddish.

Fruit: capsule, ovoid; seed with hairs.

Distribution: native to Europe, but cultivated widely in the Yangtze River basin and the Yellow River basin.

Figure 4-92 *Populus* × *canadensis* 加杨

2. *Salix* Linn. 柳属

Growth sympodial, buds with 1 scale, terminal bud absent. Leaves simple, linear-lanceolate typically with serrate. Female catkin erecting or spreading. Flowers without disc but glands sometimes connate and discoid.

Salix babylonica Linn. 垂柳 (Figure 4-93)

Habit: trees, deciduous.

Bark: grayish black, irregularly furrowed.

Twig: pendulous, brownish yellow, or slightly purple, slender, glabrous. Terminal bud absent.

Leaf: stipule, conspicuous; narrowly lanceolate or linear-lanceolate, abaxially light green, adaxially green, base cuneate, margin serrate, apex long acuminate.

Flower: inflorescence a catkin. Male flower: glands 2, stamens 2, anthers yellow. Female flower: gland adaxial, ovary ellipsoid, glabrous, style short, stigma 4-parted.

Fruit: capsule, slightly greenish brown, seed with hairs.

Distribution: native to the Yangtze River basin and the Yellow River basin, but cultivated widely in many areas of China.

Figure 4-93 *Salix babylonica* 垂柳

4.20　Myricaceae 杨梅科

Leaves alternate, simple, with entire or serrate/dentate margins, usually aromatic, glandular scales; stipules absent. Inflorescences borne in the leaf axils in catkins. Flowers unisexual, perianth absent. Male flowers spikes, stamen 2-16. Female flowers with syncarpous gynoecium, 2 stigmas, superior 1-locular ovary. Fruit a 1-seeded drupe or nutlet, wax-covered papillae; endocarp hard.

There are 1 genus and 4 species in China.

Myrica Linn. 杨梅属

The morphological characteristics are the same as the family's.

***Myrica rubra* Sieb. et Zucc. 杨梅 (Figure 4-94)**

Habit: trees evergreen, dioecious.

Bark: gray, smooth.

Twig: branchlet and bud glabrous.

Leaf: cuneate-obovate or narrowly elliptic-obovate, leathery, glabrous, abaxially pale green and sparsely to moderately golden glandular, base cuneate, margin entire or several serrates, apex obtuse to acute.

Figure 4-94 *Myrica rubra* 杨梅

Flower: male spike simple or inconspicuously branched, solitary or sometimes few together in leaf axil. Anthers dark red, ellipsoid. Female spike solitary in leaf axil. Ovary velutinous; stigmas 2, bright red, slender.

Fruit: drupe, dark red or purple-red at maturity, globose.

Distribution: south of the Yangtze River and Taiwan.

* The species is similar to *Myrica esculenta* Buch. -Ham. ex D. Don 毛杨梅, but the latter branchlet and petiole tomentose, drupe ellipsoid; distributing in southwest China, Guangdong and Guangxi.

4.21 Betulaceae 桦木科

Leaves alternate, simple. Flowers unisexual, monoecious. Male inflorescence a catkin, precocious, pendulous, with numerous overlapping bracts, containing a small dichasium with 1-3 male flowers. Female inflorescence with many small dichasium each containing 2 or 3 flowers in overlapping bracts. Fruit a nut or nutlet, winged or not.

There are 6 genera and 89 species in China.

1. *Betula* Linn. 桦木属

Male flowers 3 in a small dichasium, each male flower with 2 stamens; calyx present. Female inflorescence spicate, flowers without calyx; bracts leathery, deciduous, 3-lobed at apex, each bract subtending 3 flowers. Nutlet compressed, usually with membranous wings.

Betula platyphylla Suk. 白桦 (Figure 4-95)

Habit: trees, deciduous.

Bark: grayish white, exfoliating in sheets.

Twig: branches erect, dark gray; branchlets brown, sparsely resinous glandular.

Leaf: triangular, ovate-triangular, or broadly ovate, adaxially sparsely pubescent and resinous glandular when young, margin doubly or simply serrate, apex acute.

Flower: female inflorescence pendulous, bracts pubescent densely and ciliate, 3-lobed.

Fruit: nutlet narrowly oblong, or ovate, with membranous wings.

Distribution: from north to west in China.

* *Betula luminifera* H. Winkler 亮叶桦(光皮桦) is common in south of the Qinling Mountain-Huaihe River, its female inflorescence 1-or 2, narrowly cylindric; lateral lobes of bracts reduced; wings of nutlet partly exserted, much wider than nutlet; leave margin irregularly and doubly setiform serrate.

Figure 4-95 *Betula platyphylla* 白桦

2. *Alnus* Mill. 桤木属

Male flowers 3 in a small dichasium, flower with (1 or 3 or) 4 stamens, calyx present. Female inflorescence conelike; bracts woody, persistent, 5-lobed at apex, each bract subtending 2 flowers; flowers without calyx. Fruit a nutlet, winged; cotyledons flat.

***Alnus trabeculosa* Hand. -Mazz.** 江南桤木（Figure 4-96）

Habit: trees, deciduous.

Bark: gray or gray-brown, smooth.

Twig: gray-brown, yellow pubescent.

Leaf: obovate-oblong, oblanceolate-oblong, or oblong, abaxially resinous glandular, glabrous, base subrounded, subcordate, or broadly cuneate, margin remotely minutely serrates.

Flower: male inflorescence fascicular; female inflorescences 2-4 in a raceme, bracts, woody.

Fruit: nutlet, broadly ovate, with papery wings.

Distribution: south of the Yangtze River in China.

Figure 4-96　*Alnus trabeculosa* 江南桤木

3. *Corylus* Linn. 榛属

Male flower 1, with 2 bracteoles, calyx absent; bracts campanulate or forming a tubular sheath. Female inflorescence capitulate or racemose-capitulate, flowers with calyx adnate to ovary. Fruit a nut or nutlet, wingless.

ial *Corylus heterophylla* var. *sutchuenensis* Franch. 川榛 (Figure 4-97)

Habit: shrubs or small trees, deciduous.

Bark: gray.

Twig: pubescent and stipitate glandular, with white lenticels.

Leaf: elliptic-obovate, broadly ovate, or suborbicular, apex subrounded, mucronate. Lobes of bracts usually dentate.

Flower: male inflorescence 2-5 in a cluster, pendulous, slender; bracts reddish brown. Female flowers 2-6 in a cluster; bracts campanulate, densely pubescent and stipitate glandular near base, lobed.

Fruit: nut, ovoid-globose, apex villous.

Distribution: southwest China, north China, central China and east China.

* The variety is different from the *Corylus heterophylla* Fisch. ex Trautv. 榛, especially in leaf blade. The latter leaf blade oblong or obovate, apex mucronate or caudate to nearly truncate; lobes of bracts usually entire.

Figure 4-97 *Corylus heterophylla* var. *sutchuenensis* 川榛

4. *Carpinus* Linn. 鹅耳枥属

Male inflorescence enclosed by buds during winter, flower 1, calyx absent. Female inflorescence racemose, female flowers with calyx adnate to ovary; fruit a nut or nutlet, wingless. Fruit a nutlet, naked or seminaked.

Carpinus putoensis W. C. Cheng 普陀鹅耳枥 I (Figure 4-98)

Habit: trees, deciduous.

Bark: brown-gray.

Twig: brown, sparsely villous.

Leaf: elliptic or broadly elliptic, base rounded or broadly cuneate, margin irregularly and doublysetiform serrate, apex acute or acuminate, lateral veins 11-14 on each side of midvein.

Flower: female inflorescence bracts semiovate.

Fruit: nutlet, broadly ovoid, glabrous, except villous at apex, sometimes with sparse glands, prominently ribbed.

Distribution: Zhoushan Islands, Zhejiang province.

Figure 4-98 *Carpinus putoensis* 普陀鹅耳枥

Carpinus turczaninowii Hance 鹅耳枥 (Figure 4-99)

Habit: trees, deciduous.

Bark: dark gray.

Twig: gray-brown, slender, pubescent when young, glabrescent.

Leaf: ovate, broadly ovate, ovate-elliptic, bearded in axils of lateral veins, base subrounded or broadly cuneate, margin regularly or irregularly doubly serrate, apex acute or acuminate.

Flower: female inflorescence bracts semiovate, semioblong, or broadly semiovate.

Fruit: nutlet, broadly ovoid, glabrous except sparsely villous at apex, resinous glandular, prominently ribbed.

Distribution: north China, Gansu, Shaanxi and Jiangsu (Yuntai Moutain).

＊*Carpinus oblongifolia* (Hu) Hu & W. C. Cheng 宝华鹅耳枥 is native to Baohua Mountain, Jiangsu Province, whose key features: leaf blade obovate-oblong, margin irregularly and doubly serrate; bracts semiovate, inner margin entire or remotely and obscurely serrate, with inflexed basal auricle. Nutlet broadly ovoid.

Figure 4-99 *Carpinus turczaninowii* 鹅耳枥

4.22 Fagaceae 壳斗科

Leaves alternate, simple, lobed, toothed, or entire margins. Flowers unisexual (monoceious); usually 6 reduced/inconspicuous tepals; 4 to many stamens; 3(-12) carpels, connate in an inferior ovary. Male inflorescences in dangling catkins. Female inflorescences in sessile clusters. Fruit a nut, associated with a spiny or scaly cupule.

There are 7 genera and 294 species in China.

1. *Fagus* Linn. 水青冈属

Male inflorescence head, pendulou; female flowers (1 or) 2. Cupules woody, (3 or) 4 - valved; bracts leaflike or filiform to short, triangular. Nuts 2 usually, ovoid to 3 - sided. Germination epigeal.

***Fagus engleriana* Seem.** 米心水青冈（Figure 4-100）

Habit: trees, deciduous.

Bark: strikingly smooth and dark grey, but slightly furrowed when mature.

Twig: shiny brown.

Leaf: ovate, elliptic-ovate, abaxially glaucescent and glabrous except for long silky hairs along veins, base broadly cuneate, margin sinuate, apex shortly acuminate.

Flower: male inflorescence in leaf axils toward base of branchlets, a lax pedunculate head; perianth 4-7-lobed. Female inflorescence in axils of leaves; flowers usually 2, subtended by a single (3-or) 4-parted cupule and surrounded by numerous bracts.

Fruit: nut, slightly exserted, apex with 3 small wings.

Distribution: east China, central China, southwest China and Shaanxi.

* The species is similar to *Fagus longipetiolata* Seem. 水青冈, but the latter: cupules covered with filiform and recurved or short, 2-2.5 cm, triangular bracts; leaf blade secondary veins ending in teeth.

Figure 4-100 *Fagus engleriana* 米心水青冈

2. *Castanea* Mill. 栗属

Trees, deciduous. Axillary buds of most apical leaves in false-terminal buds. Male inflorescence erect catkin; stamens 10-12 in each male flower. Ovary inferior, 6-9-loculed; stigma terminal, minutely punctiform. Cupule splitting into 2-4 valves; bracts spinelike. Fruit a nut, 1-3 per cupule. Germination hypogeal.

Castanea mollissima **Blume** 板栗（Figure 4-101）

Habit: trees, deciduous.

Bark: dark gray; strongly ridged and furrowed when mature.

Twig: short pubescence, often also with long spreading hairs.

Leaf: elliptic-oblong to oblong-lanceolate, abaxially tomentose to softly pubescent, adaxially scalelike glands sometimes absent, base rounded, margin coarsely serrate, apex acute.

Flower: male inflorescence a catkin, erect, smelly. Cupule densely covered with pubescent spinelike bracts.

Fruit: nut, 2 or 3 per cupule usually.

Distribution: cultivated in many areas of China.

Figure 4-101 *Castanea mollissima* 板栗

Castanea henryi **(Skan) Rehd. et E. H. Wils.** 锥栗（Figure 4-102）

Habit: trees, deciduous.

Bark: whiteish gray, furrowed when mature.

Twig: dark purple brown, glabrous.

Leaf: lanceolate to ovate lanceolate, abaxially covered with yellowish brown scalelike glands and sparsely pilose along veins when young, glabrescent, base rounded to broadly cuneate, margin with bristlelike teeth.

Flower: male inflorescence a catkin, erect. Female flowers 1(-3) per cupule. Cupule on a short spike, covered with slightly pubescent spinelike bracts.

Fruit: nut 1 per cupule, globose-ovoid.

Distribution: south of the Qinling Mountains-Huaihe River in China.

* *Castanea seguinii* Dode 茅栗 is also common in field, but abaxially covered with scalelike glands, glabrous; nuts usually 3 per cupule, usually shorter than the wild.

Figure 4-102 *Castanea henryi* 锥栗

3. *Castanopsis* (D. Don) Spach 栲属(锥属)

Trees, evergreen. Terminal buds conspicuous. Leaves alternate, distichous, base unequal. Male inflorescence elongated catkins, erect or pendulous; female flowers solitary or in spikes, ovary 3-loculed. Cupule solitary on rachis, different shap or spine. Fruit a nut, 1-3 per cupule. Germination hypogeal.

Castanopsis sclerophylla (Lindl. et Paxton) Schottky 苦槠 (Figure 4-103)

Habit: trees, evergreen.

Bark: dark gray, longitudinally fissured.

Twig: reddish brown, glabrous, slightly angulate when young.

Leaf: distichous. Leaf blades oblong, ovate-elliptic, leathery, adaxially silver-gray with age, base inaequilateral usually, margin from middle to apex serrulate or rarely entire, apex acuminate, cuspidate, or shortly caudate.

Flower: cupule globose, completely or almost completely enclosing nut, irregularly valved; bracts scalelike, 3- or 4-angled, sometimes only base connate, in annular umbones.

Fruit: nut 1(-3) per cupule, subglobose, tomentulose, apex mucronulate.

Distribution: south of the Qinling Mountains-Huaihe River, but concentrated in east China.

Figure 4-103 *Castanopsis sclerophylla* 苦槠

Castanopsis eyrei (Champ. ex Benth.) Tutcher 甜槠 (Figure 4-104)

Habit: trees, evergreen.

Bark: greyish brown, longitudinally fissured.

Twig: glabrous.

Leaf: leathery, ovate to ovate-lanceolate, rarely ovate-elliptic, base cuneate, ±oblique, apex caudate-acuminate, margin entire or with a few coarse serrates towards apex, green, glabrous and shining on both surfaces.

Flower: female spike simple and shorter; flower 1 per cupule. Cupule subglobose to broad ovoid, splitting irregularly when mature; bracts spine-like, usually entirely covering cupule.

Fruit: nut 1 per cupule, broadly conical.

Distribution: south of the Yangtze River basin, except Yunnan and Hainan provinces.

Figure 4-104　*Castanopsis eyrei* 甜槠

4. *Lithocarpus* Blume 柯属(石栎属)

Trees, evergreen. Male inflorescence elongated catkins, erect; female flower solitary, ovary 3-loculed. Cupules in cymes on rachis. Germination hypogeal.

Lithocarpus glaber (**Thunb.**) **Nakai** 石栎(柯) (**Figure 4-105**)

Habit: trees, evergreen.

Bark: blackish-brown and smooth at first, becoming furrowed and rough with age.

Twig: densely tawny tomentose.

Leaf: wide elliptical to oblong, abaxially with dense scalelike glands, margin entire or with inconspicuous teeth near the apex, base cuneate.

Flower: male inflorescence in a panicle or solitary in leaf axils. Female inflorescence often with a few male flowers; cupule in clusters of 3(-5). Cupule plate-to-cupular; bracts imbricate or connate into concentric rings, triangular, small, appressed, densely puberulent.

Fruit: nut ellipsoid, apex pointed.

Distribution: from east China, central China to south China.

Figure 4-105 *Lithocarpus glaber* 石栎

5. *Cyclobalanopsis* Oersted 青冈属

Trees, evergreen. Male inflorescence catkin, pendulous; stamens 6 usually. Female inflorescences with a solitary flower or spiciform; cupule with 1 flower; perianth 5- or 6-lobed; stigmas capitate or dilated. Cupule solitary; bracts scalelike, whorled, connate, in rings or spirally fused. Nut 1 per cupule usually. Germination hypogeal.

Cyclobalanopsis glauca (Thunb.) Oersted 青冈 (Figure 4-106)

Habit: trees, evergreen.

Bark: grayish green, smooth.

Twig: glabrous.

Leaf: elliptic, oblong, or narrow-ovate, apex acuminate, base round, glabrous above, glaucous underside, margins of leaf sharply toothed in the upper part.

Flower: monoecious, yellowish-green flowers in separate male and female catkins. Cupule bowl-shaped, enclosing 1/3 – 1/2 of nut; bracts in 5 or 6 rings, crowded, margin entire or denticulate.

Fruit: nut ovoid, oblong-ovoid, or ellipsoid, glabrous.

Distribution: south of the Yangtze River basin, Shaanxi and Taiwan.

* the species is similar to *Cyclobalanopsis gracilis* (Rehd. et E. H. Wils.) W. C. Cheng et T. Hong 细叶青冈, but the latter leaf blade narrower, ca. 1.5-2.5 cm wide.

Figure 4-106 *Cyclobalanopsis glauca* 青冈

6. *Quercus* Linn. 栎属

Terminal bud conspicuous. Male inflorescence catkin, pendulous; stamens 6 usually. Female flower solitary, stigmas dilated or ligulate, lining inner faces of styles. Cupule solitary; bracts imbricate, not united, arranged spirally, scalelike, linear, or conical, adherent, prostrate, or reflexed. Nut 1 per cupule. Germination hypogeal.

Quercus acutissima Carr. 麻栎（Figure 4-107）

Habit: trees, deciduous.

Bark: dark gray, deeply furrowed.

Twig: yellowish gray tomentose, glabrescent, lenticellate.

Leaf: narrowly elliptic-lanceolate, concolorous, tomentose, glabrescent or only veins abaxially tomentose with age, margin with spiniform teeth; secondary veins fusing atserration.

Flower: cupules on previous year's branchlets, 1 or 2, cupular to discoid. Bracts, enclosing 1/4-1/2 of nut, bracts subulate to ligulate, reflexed.

Fruit: nut, ovoid to ellipsoid.

Distribution: from south of Liaoning to northwest China.

* the species is similar to *Quercus variabilis* Bl. 栓皮栎 and *Quercus chenii* Nakai 小叶栎, but they can be distinguished by leaf blades and branchlets. *Q. variabilis* mature leaf blades densely grayish stellate tomentose, and branchlets glabrous. *Q. chenii* mature leaf blades smaller than the other two, glabrous or pubescent only abaxially along veins, young branchlets pubescent.

Figure 4-107 *Quercus acutissima* 麻栎

Quercus dentata Thunb. 槲树（Figure 4-108）

Habit: trees, deciduous.

Bark: dark grey, rough, later with deep grooves.

Twig: strong, sulcate, densely yellowish gray stellate tomentose.

Leaf: petiole short or sessile, leaf blade obovate to narrowly, abaxially densely grayish brownstellate tomentose, base rounded, margin with a few undulate to rough serrations on each side, apex with short, blunt tip.

Flower: cupule cupular, enclosing 1/2 – 2/3 of nut; bracts reddish brown, narrowly lanceolate, inflexed or erect, leathery, abaxially with brown filiform hairs.

Fruit: nut ovoid to broadly.

Distribution: widely in China, especially in east and north parts.

Figure 4-108 *Quercus dentata* 槲树

Quercus serrata Murr. 枹栎（Figure 4-109）

Habit: trees, deciduous.

Bark: dark grey, or whiteish grey; smooth when young, but deep grooves later.

Twig: slender with grooves, pubescent but glabrescent.

Leaf: subsessile to petiolate, leaf blade narrowly elliptic-ovate, ovate-lanceolate, or obovate, abaxially glabrous or occasionally stellate tomentose, base cuneate to nearly rounded, margin glandular serrate, apex acuminate to acute.

Flower: cupule cupular, enclosing 1/4 - 1/3 of nut; bracts triangular, adherent, margin pilose.

Fruit: nut, ovoid to ovoid-globose.

Distribution: south of the Yangtze River basin, north China, Gansu, Shaanxi, Liaoning and Taiwan.

Figure 4-109　*Quercus serrata* 枹栎

Quercus phillyreoides A. Gray 乌冈栎 (Figure 4-110)

Habit: trees, evergreen.

Bark: brown, glabrous.

Twig: slender, grayish brown pubescent, gradually glabrescent.

Leaf: obovate, narrowly elliptic, base rounded to nearly cordate, margin glandular serrulate, apex mucronate to shortly acuminate; tertiary veins abaxially inconspicuous to evident but very slender.

Flower: cupule cupular, enclosing 1/3-1/2 of nut; bracts crowded, grayish pubescent.
Fruit: nut ellipsoid.
Distribution: from south of the Qinling Mountains to north part of south China.

Figure 4-110 *Quercus phillyreoides* 乌冈栎

4.23 Juglandaceae 胡桃科

Resinous, aromatic trees with alternate, pinnately divided leaves. Branchlets with solid or chambered pith. Flowers unisexual, monoecious. Male flowers catkins, with 3 to 6 sepals. Female flowers 4 sepals. Pistil syncarpous with 2 to 3 carpels fused together to make a single-chambered ovary. Fruit a drupelike nut or a 2- or 3-winged or disc-winged nutlet.

There are 7 genera and 20 species in China.

1. *Juglans* Linn. 胡桃属

Branchlet with chambered pith. Male flowering spike pendulous; female spike erect, perianth 4-lobed. Fruit a large, drupelike nut, indehiscent or with an irregularly dehiscent husk; germination hypogeal.

Juglans regia **Linn.** 核桃(胡桃)(**Figure 4-111**)

Habit: trees, deciduous.

Bark: ashey gray, with flattened ridges, developing a striking diamond shaped pattern.

Twig: stout, light brown, with a buff-colored chambered pith; terminal buds large, broadly pointed.

Leaf: alternate, pinnately compound with 5 to 13 leaflets; leaflets ovate to obovate with entire or sometimes finely serrated margins, terminal leaflets the largest; very aromatic when crushed.

Flower: monoecious; males single-stemmed catkins; females in clusters of 3-9, with or just after the leaves.

Fruit: nut, subglobose; husk glabrous, irregularly dehiscent; shell thick, wrinkled.

Distribution: central China, east China, northwest China, and southwest China; cultivated widely now.

Figure 4-111 *Juglans regia* 核桃

***Juglans mandshurica* Maxim.** 核桃楸(野核桃)(**Figure 4-112**)

Habit: trees, deciduous.

Bark: greyish white or dark grey, rough, deep grooves.

Twig: stout, greyish brown. Terminal buds with false-valved scales.

Leaf: petiole and rachis sparsely to moderately glandular pubescent. Leaflets 9-19, lateral ones sessile, blade elliptic to long elliptic, abaxially tomentose, base oblique, subcordate, margin serrate, rarely serrulate, apex acuminate.

Flower: male spike 9-40 cm. Stamens 12-40.

Fruit: nut, globose, ovoid, or ellipsoid; husk densely glandular pubescent, indehiscent; shell thick, rough, with prominent ridges and deep pits and depressions.

Distribution: east China, southwest China, northeast China, Gansu and Shaanxi.

Figure 4-112　*Juglans mandshurica* 核桃楸

2. *Pterocarya* Kunth 枫杨属

Branchlet with chambered pith. Fruiting spike elongate, pendulous. Fruit a 2-winged nutlet.

***Pterocarya stenoptera* C. de Candolle** 枫杨（**Figure 4-113**）

Habit: trees, deciduous.

Bark: dark grey, deep grooves.

Twig: stout. Lenticels yellow. Terminal buds naked, with brown glandular scale.

Leaf: even-pinnate, rachis often winged; leaflets sessile, long elliptic to elliptic-lanceolate, abaxially slightly pubescent, base oblique, cuneate or broadly cuneate, apex obtuse or acute.

Flower: male flower with an entire bract. Female flower with a small, entire bract, adnate to ovary; sepals 4, adnate to ovary, free at apex; style short; stigmas carinal, 2-lobed, plumose.

Fruit: nutlets long ellipsoid, wings linear.

Distribution: from east China to southwest China, Shaanxi and Henan.

Figure 4-113 *Pterocarya stenoptera* 枫杨

3. *Cyclocarya* Iljinsk. 青钱柳属

Branchlet with chambered pith. Leaves imparipinnate. Inflorescence a catkin. Fruit a disc-winged nutlet.

Cyclocarya paliurus (Batalin) Iljinsk. 青钱柳(Figure 4-114)

Habit: trees, deciduous.

Bark: grey, fibrous, fissured when older.

Twig: Branchlet with chambered pith. Terminal buds naked.

Leaf: imparipinnate, rachis tomentose, not winged; leaflets 7-9, sessile, elliptic-ovate to broadly lanceolate, base oblique, broadly cuneate to subrounded, apex obtuse or acute, rarely acuminate, margin finely serrate.

Flower: male spike 8-13 cm, in clusters of 3-5, lateral on old growth, pendulous. Female spike solitary, 25-30 cm, terminal on new growth.

Fruit: nutlets compressed globose, small; disc wing leathery, orbicular to ovate.

Distribution: south of the Yangtze River basin and Taiwan.

Figure 4-114　*Cyclocarya paliurus* 青钱柳

4. *Carya* Nutt. 山核桃属

Pith solid and homogeneous. Leaves odd-pinnate. Male spikes in clusters of 3, female spike terminal on new growth. Fruit a drupelike nut with a thick, 4-valved husk covering a smooth or wrinkled shell 2-4-chambered at base. Germination hypogeal.

Carya illinoensis (Wangen.) K. Koch 美国山核桃 (Figure 4-115)

Habit: trees, deciduous.

Bark: light gray or brownish, ridged with appressed scales or exfoliating with small platelike scales.

Twig: reddish brown, slender, hirsute, conspicuously scaly, sometimes becoming glabrous. Terminal buds yellowish brown, hirsute, scaly.

Leaf: large; rachis generally glabrous or glabrescent; leaflets 9-13, lateral ones shortly petiolulate or sessile, blade ovate-lanceolate to elliptic-lanceolate, with scattered, peltate scales, base oblique, apex acuminate, margin finely serrate.

Flower: staminate catkins essentially sessile, to 18 cm, stalks with small capitate-glandular trichomes; anthers sparsely pilose.

Fruit: dark brown, ovoid-ellipsoid, not compressed; husks rough, sutures winged; nuts tan to brown and mottled with black patches, ovoid-ellipsoid, not compressed, smooth; shells thin.

Distribution: native to the United States, but cultivated in east China and other regions.

Figure 4-115 *Carya illinoensis* 美国山核桃

5. *Platycarya* Sieb. et Zucc. 化香树属

Branchlet with solid pith. Leaves imparipinnate. Inflorescences clusters of erect catkins terminal on new growth; female cone-like with rigid, persistent bracts. Fruit a small, flattened, narrowly winged nutlet, 2-chambered at the base. Germination epigeal.

Platycarya strobilacea Sieb. et Zucc. 化香树 (Figure 4-116)

Habit: small trees, deciduous.

Bark: dark grey, rough and fissured.

Twig: pubescent at first, becoming glabrous, yellow-brown to chestnut, with solid pith.

Leaf: imparipinnate, glabrous. Lateral leaflets sessile, blade ovate-lanceolate to narrowly elliptic-lanceolate, base oblique to cuneate, lower surfaces glabrous except for dense cluster of hairs at base and along midvein.

Flower: erect catkin, clustered at the tips of new growth. Androgynous spikes, yellowish female flowers in broad, cone-like basal portion; male spikes, yellowish to bright yellow and scented early in fertile period.

Fruit: fruiting spike cone-like, with persistent bracts, ovoid-ellipsoid or ellipsoid-cylindric to subglobose. Nuts flattened, narrowly two-winged.

Distribution: east China, central China, south China and southwest China.

Figure 4-116 *Platycarya strobilacea* 化香树

4.24 Ulmaceae 榆科

　　Leaves simple, alternate, base asymmetrical slightly. Flowers bisexual or unisexual; separate sepals 4 or 8, without petal; stamens 4 to 8. Ovary positioned superior with 2 united carpels (bicarpellate), forming a single chamber. Fruit a samara, drupe, or winged nutlet, apically usuallywith persistent stigmas.

　　There are 8 genera and 46 species in China.

1. *Ulmus* Linn. 榆属

　　Leaves venation pinnate; secondary veins extending to margin. Fruit a samara flat, usually more or less circular, with the seed in the center or towards the apex.

***Ulmus pumila* Linn.** 榆树 (**Figure 4-117**)

　　Habit: trees, deciduous.
　　Bark: dark gray, irregularly longitudinally fissured.

Figure 4-117 *Ulmus pumila* 榆树

Twig: yellowish gray, glabrous or pubescent, unwinged and without a corky layer, with scattered lenticels.

Leaf: oval or ovate-lanceolate, base obliquely to symmetrically obtuse to rounded, margin simply or sometimes doubly serrate, apex acute to acuminate.

Flower: inflorescence a fascicled cyme in the 2nd year's branchlets. Perianth 4-lobed, margin ciliate.

Fruit: samaras circular or rather obovate.

Distribution: commonly in north China.

Ulmus parvifolia Jacq. 榔榆 (Figure 4-118)

Habit: trees, deciduous.

Bark: gray to grayish brown, exfoliating into irregular scale-like flakes.

Twig: slender, dark brown, densely pubescent when young.

Leaf: simple, lanceolate-ovate to narrowly elliptic, base oblique, margin obtusely and irregularly simply serrate, apex acute to obtuse.

Flower: inflorescence a fascicled cyme. Perianth funnel form; tepals 4.

Fruit: samara, tan to brown, elliptic to ovate-elliptic; seed in the center.

Distribution: east China, north China, northwest China, south China, Shaanxi and Henan.

Figure 4-118 *Ulmus parvifolia* 榔榆

2. *Zelkova* Spach 榉属

Leaves distichous, margin serrate to crenate; venation pinnate; secondary veins extending to margin. Fruit a drupe, oblique, dorsally keeled; endocarp hard; perianth persistent; stigmas beak-shaped.

***Zelkova schneideriana* Hand. -Mazz.** 大叶榉树 Ⅱ（Figure 4-119）

　　Habit：trees, deciduous.

　　Bark：purplish red to grey, scaly and peeling.

　　Twig：grey or greyish brown, covered with whitish pubescence.

　　Leaf：thickly papery, ovate to elliptic-lanceolate, upper surface strigose, lower surface green to reddish purple and densely pubescent, margin serrate to crenate, apex acuminate to acute.

　　Flower：staminate flower solitary or in clusters of 2 to 3; pistillate and hermaphrodite flowers usually solitary.

　　Fruit：drupe, pea-green, subsessile, covered in an irregular network of ridges.

　　Distribution：south of the Qinling Mountains-Huaihe River, Gansu and Henan.

　　* The species is similar to *Zelkova sinica* C. K. Schneid. 大果榉, but the latter drupe larger (5-7 mm in diam.), obovoid-globose, surface smooth, apex only slightly oblique.

Figure 4-119　*Zelkova schneideriana* 大叶榉树

3. *Celtis* Linn. 朴属

Leaf blade margin entire or serrate; 3-veined from base; secondary veins anastomosing beforereaching margin. Fruit a drupe; embryo curved; cotyledons broad.

Celtis sinensis **Pers.** 朴树 (Figure 4-120)

Habit: trees, deciduous.

Bark: light gray, smooth.

Twig: some brown pubescence in the 1st year, but glabrous later.

Leaf: elliptical to ovate, papery, margins subentire to crenate, apex acute to acuminate, secondary veins 3 or 4 on each side of midvein.

Flower: densely clustered cyme in leaf axil and at base of young stems.

Fruit: drupe, globose. Infructescences unbranched, stout, pubescent. Fruit a drupe globose.

Distribution: east China, central China, southwest China, Henan, Guangdong.

* *Celtis julianae* C. K. Schneid. 珊瑚朴 is also common in field, whose features are petiole and abaxially golden pubescent, and drupe larger.

Figure 4-120 *Celtis sinensis* 朴树

4. *Aphananthe* Planch. 糙叶树属

Leaf margin serrate or entire, 3-veined, secondary veins less 6 pairs, extending to margin. Drupes ovoid to ± globose; cotyledons narrow.

Aphananthe aspera (**Thunb.**) **Planch.** 糙叶树 (**Figure 4-121**)

Habit: trees, deciduous.

Bark: brown or grayish brown, scabrous, longitudinally fissured.

Twig: yellowish green when young, brownish red in the 2nd year; the old one grayish brown, with distinct rounded lenticels.

Leaf: alternate, ovate, base wedge-shaped, rounded, often oblique, prominently parallel-veined, distinctly 3-nerved at the base, both surfaces densely covered with minute, flattened hairs.

Flower: unisexual, very small, greenish; males numerous, crowded in slender, stalked, cymose clusters; females solitary.

Fruit: drupe, roundish oval, black-purple.

Distribution: north China, east China, central China, south China, southwest China.

Figure 4-121 *Aphananthe aspera* 糙叶树

5. *Pteroceltis* Maxim. 青檀属

Leaves often distichous, blade serrate, 3-veined from base; secondary veins branching and often anastomosing before reaching marginal teeth. Anther apically pubescent. Nut broadly winged.

***Pteroceltis tatarinowii* Maxim.** 青檀（Figure 4-122）

Habit: trees, deciduous.

Bark: grayish white to dark gray, exfoliates in patches.

Twig: slender, with distinct lenticels.

Leaf: broadly ovate to oblong, base oblique, margin irregularly serrate, apex acuminate, secondary veins 4-6 on each side of midvein.

Flower: unisexual; male in stalkless clusters, females solitary in the leaf-axils.

Fruit: nut, globose, surrounded by a circular wing notched at the top.

Distribution: north China, east China, central China, south China, southwest China, Liaoning, Gansu, Shaanxi and Qinghai.

Figure 4-122   *Pteroceltis tatarinowii* 青檀

4.25   Moraceae 桑科

Milky or watery latex in branchlets usually. Leaves alternate. Flowers small, unisexual (monoecious or dioecious). Ovary superior, semi-inferior, or inferior, 1-loculed. Fruit a drupe or achene, often coalesced or otherwise aggregated into a multiple accessory fruit.

There are 9 genera and 114 species in China.

1. *Maclura* Nutt. 橙桑属(柘属)

Trees or small trees, with latex; dioecious. Spine usually present. Stipules free. Leaf margin entire. Inflorescence axillary, globose. Filaments erect. Fruit a drupelet, ovoid, surface shell-like, enveloped by a fleshy calyx.

Maclura tricuspidata Carr. 柘 (Figure 4-123)

Habit: small trees or shrubs, deciduous.

Bark: grayish brown, ripples with deep furrows.

Twig: glabrous, immature wood thorny.

Leaf: alternate, oval, obovate, or ovate, either entire or with three shallow rounded lobes, at the apex.

Flower: inflorescence a head-like; dioecious; flower green, pea-sized.

Fruit: aggregate fruit, knotty, ripens to red or maroon-red, juicy, rich flesh.

Distribution: east China, central China, and southwest China (reach to Shaanxi and Hebei).

Figure 4-123 *Maclura tricuspidata* 柘

2. *Morus* Linn. 桑属

Winter bud with 3-6 bud scales. Stipules free, caducous. Leaves blade simple to deeply palmately lobed, margin toothed. Monoecious or dioecious. Inflorescence a catkin, axillary. Fruit with enlarged, succulent calyx usually aggregated into juicy syncarp.

***Morus alba* Linn.** 桑 (Figure 4-124)

　　Habit: trees, deciduous.
　　Bark: gray, shallowly furrowed.
　　Twig: downy, becoming glabrous later.
　　Leaf: simple, ovate to broadly ovate, irregularly lobed, base rounded to ± cordate, margin coarsely serrate to crenate, apex acute, acuminate, or obtuse.
　　Flower: male and female catkins pendulous, axillary. Style inconspicuous or absent; stigmas 2.
　　Fruit: syncarp red when immature, blackish purple, purple, or greenish white when mature.
　　Distribution: central China and north China, now cultivated all over China.

Figure 4-124　*Morus alba* 桑

3. *Broussonetia* L'Héritier ex Vent. 构属

Winter bud small, with scales 2-3. Stipules free, caducous. Leaves blade simple to palmately lobed, margin toothed. Male inflorescence a catkin, pendulous; female inflorescence a densely capitate. Fruit densely aggregated into globose syncarp.

Broussonetia papyrifera (Linn.) L'Héritier ex Vent. 构树 (Figure 4-125)

Habit: trees, deciduous.
Bark: dark gray, smooth or shallowly grooved.
Twig: thickly downy, soft and pithy.
Leaf: variable in size and form, ovate or variously lobed, rounded, or more or less tapered at the base, pointed, toothed, 3-nerved at the base, upper surface dull green and rough, lower surface densely woolly.
Flower: male plant in cylindrical often curly, woolly catkins; female flowers in ball-like heads.
Fruit: syncarp orange-red when mature.
Distribution: almost all over China.

* the species is similar to *Broussonetia kazinoki* Sieb. 小构树(楮), but the latter: shrubs; twigs slender, climbing, with tomentose; leaf blades ovate to oblique-ovate; monoecious; syncarp small, 5-6 mm, freshy, red when mature.

Figure 4-125 *Broussonetia papyrifera* 构树

4. *Ficus* Linn. 榕属

Latex conspicuous in plant. Stipules connate, lateral to amplexicaul and enclosing terminal leaf bud, scar ringlike. Leaves usually alternate, margin entire. Inflorescence a unique structure known as a syconium – a hollow structure containing minute flowers. Fruit a seedlike achene, usually enclosed within syncarp formed from an enlarged hollow fleshy receptacle.

***Ficus carica* Linn.** 无花果 (Figure 4-126)

Habit: shrubs, deciduous.

Bark: greyish white, smooth.

Twig: stout, grayish brown, distinctly lenticellate.

Leaf: stipule red, large. Leaf blade broadly ovate, usually with 3 – 5 ovate lobes, thickly papery, margin irregularly toothed; basal lateral veins 2-4.

Flower: produced on the inner surface of a roundish, pear-shaped receptacle, nearly closed at the top, which afterwards develops into the succulent sweet fruit, known as the fig.

Fruit: syconus, fig; achene lenslike.

Distribution: native to Mediterranean, but cultivated throughout China.

Figure 4-126 *Ficus carica* 无花果

4.26 Eucommiaceae 杜仲科

Trees deciduous. Branchlets and leaves containing latex. Flower unisexual, dioecious, without perianth. Fruit an indehiscent samara with one seed, narrowly winged, containing latex.

There are 1 genera and 1 species in this family, endemic to China.

Eucommia Oliv. 杜仲属

Morphological characters are the same as those of family.

Eucommia ulmoides Oliv. 杜仲 (**Figure 4-127**)

Habit: trees, deciduous.

Bark: gray-brown, scabrous.

Figure 4-127 *Eucommia ulmoides* 杜仲

Twig: yellow-brown pubescent at first, soon glabrate, old ones conspicuously lenticellate. Bud shiny red-brown.

Leaf: alternate, ovate or oval, long and slender-pointed, toothed, slightly hairy on both surfaces when young, becoming glabrous above later.

Flower: unisexual, dioecious; small and inconspicuous, the males consisting of brown stamens only; female ones of a single pistil.

Fruit: samara, flat and winged, one-seeded.

Distribution: northwest China, central China and southwest China.

4.27　Pittosporaceae 海桐花科

Trees or shrubs, evergreen. Stem resin canal. Leaves alternate, estipulate; mostly leathery, margin entire. Inflorescence umbellate, corymbose, paniculate, or a solitary flower, bracteate and bracteolate. Flowers usually bisexual. Petals free or connate. Fruit a capsule dehiscing by adaxial suture, or a berry. Seeds numerous; testa thin; endosperm well developed.

There are 1 genus and 46 species in China.

Pittosporum Banks ex Gaertn. 海桐花属

Leaves alternate, appearing opposite or pseudoverticillate, usually clustered at branchlet apex; leathery, margin entire. Flowers bisexual. Sepals 5, free. Petals 5, free or partly connate, reflexed. Capsule ellipsoid or globose, dehiscing by 2-4 valves; pericarp woody or leathery.

Pittosporum tobira (Thunb.) W. T. Ait. 海桐 (Figure 4-128)

Habit: small trees orshrubs, evergreen.

Bark: brownish gray.

Twig: dark grey, stout, lenticellate.

Leaf: clustered at branchlet apex; leaf blade dark green and shiny adaxially, obovate, leathery, base narrowly cuneate, margin entire, revolute, apex rounded or obtuse.

Flower: inflorescence a terminal or near so, umbellate or corymbose; flowers fragrant, yellowish green. Sepals, petals, stamens, each 5.

Fruit: capsule, globose, angular, dehiscing by 3 valves.

Distribution: south of the Yangtze River.

＊ the species is similar to *Pittosporum illicioides* Mak. 海金子(崖花海桐), but the latter leaf blade apex obtuse or acuminate.

Figure 4-128 *Pittosporum tobira* 海桐

4.28 Tiliaceae 椴树科

  Stellate hair conspicuous usually. Bark containing fibre abundant. Sepals 5, free or sometimes basally connate. Stamens numerous, rarely 5, free or connate into fascicles at base; anthers 2-loculed. Fruit usually a drupe, capsule, or schizocarp, sometimes a berry or samara, 2-10-loculed.

  There are 11 genera and 70 species in China.

1. *Tilia* Linn. 椴树属

  Trees, deciduous. Floral bracts narrowly rectangular, large and persistent, and fused at least to the base of the inflorescence peduncle. Sepals 5, with adaxial nectary at base. Fruit a nut or capsule.

Tilia miqueliana Maxim. 南京椴（Figure 4-129）

Habit: trees, deciduous.

Bark: dull grey, developing close, lumpy and corky ridges.

Twig: quite slender, usually with dense white stellate hairs.

Leaf: rather triangular and broadest near the base, usually cordate, marginal teeth quite long and usually forward-pointing, under leaf grey with stellate pubescence.

Flower: floral bract sessile, densely covered beneath with grey-white stellate hairs. Inflorescence drooping, cup-shaped flowers. Staminodes present.

Fruit: spherical, often weakly mamillate, with dense stellate hairs and a thick wall.

Distribution: east China.

* *Tilia henryana* var. *subglabra* V. Engler 糯米椴 is common in field, but its branchlets and buds glabrous or nearly so, leaf blade abaxially hairy in vein axils only, marginal teeth 3-5 mm, bracts adaxially glabrous.

Figure 4-129 *Tilia miqueliana* 南京椴

2. *Grewia* Linn. 扁担杆属

Leaves basal veins 3-5, margin serrate or rarely lobed. Stamens normal, anthers globose. Fruit a drupe usually with 2 or 4 drupelets.

Grewia biloba G. Don 扁担杆

Habit: shrubs or small trees, deciduous.

Twig: pubescent or nearly glabrous.

Leaf: ovate-orbicular or obovate-elliptic, sometimes shallowly 3-lobed, distinctly three-nerved at the base, abaxially sparsely stellate hairy to stellate tomentose, base cuneate or obtuse, margin serrulate, apex acute.

Flower: small, yellowish green. inflorescence a cyme axillary.

Fruit: drupe red; drupelets 1 per lobe.

Distribution: east China and south China.

* The species is similar to *Grewia biloba* var. *parviflora* (Bunge) Hand. -Mazz. 小花扁担杆(Figure 4-130), but the latter leaf blade densely softly stellate tomentose abaxially, flowers smaller.

Figure 4-130 *Grewia biloba* var. *parviflora* 小花扁担杆

4.29 Elaeocarpaceae 杜英科

Leaves simple. Inflorescence axillary or terminal, racemose, corymbose, paniculate, or sometimes fascicled or solitary flowers. Flowers bisexual or polygamous, 4 - or 5 - merous, actinomorphic. Petals 4 or 5, sometimes absent, valvate or imbricate, margin laciniate. Disk circular or glandularly lobed. Fruit a drupe or capsule.

There are 2 genera and 53 species in China.

Elaeocarpus Linn. 杜英属

Inflorescence racemose; petal margin laciniate. Fruit a drupe.
Elaeocarpus glabripetalus Merr. 秃瓣杜英 (Figure 4-131)
Habit: trees, evergreen.

Figure 4-131 *Elaeocarpus glabripetalus* 秃瓣杜英

Bark: grey, smooth.

Twig: red-brown when dry, ± angular.

Leaf: oblanceolate, papery or membranous, glabrous, base narrow and decurrent, margin minutely crenate, apex acute, acumen obtuse.

Flower: inflorescence a raceme. Sepals 5, lanceolate. Petals 5, white. Filaments very short; anthers fascicled, not awned but pubescent at apices. Disk 5-lobed, pubescent.

Fruit: drupe, ellipsoid; endocarp thinly bony.

Distribution: east China, south China and southwest China.

4.30　Sterculiaceae 梧桐科

Stellately hairy usually in plant. Sepals 3-5, ± connate. Petals 5 or lacking. Androgynophore usually present; stamens many, filaments usually connate into a single tube; staminodes 5. Fruit usually a capsule or follicle, dehiscent or indehiscent.

There are 19 genera and 90 species in China.

1. *Firmiana* Marsili 梧桐属

Leaves simple, palmately 3-5-lobed, alternate. Inflorescence paniculate or rarely racemose, axillary or terminal. Fruit a capsule, follicle-like, endocarp membranous, dehiscent long before maturity, foliaceous.

Firmiana simplex (Linn.) W. Wight 梧桐 (Figure 4-132)

Habit: trees, deciduous.

Bark: greenish, smooth.

Twig: Stout, green, large nearly circular leaf scar; bud large, round with numerous fuzzy red-brown scales.

Leaf: alternate, simple, very large 3 to 5 lobed leaves, very long petiole, bright green above, often fuzzy beneath.

Flower: large, upright, loose, terminal clusters of yellow-green flowers, appearing in mid-summer.

Fruit: capsule, follicle-like, membranous.

Distribution: from Hainan Island to north China.

2. *Reevesia* Lindl. 梭罗树属

Inflorescence paniculate, cymose or thyrsoid, terminal. Flowers bisexual. Fruit a capsule, woody, 5-valves. Seeds 2 per locule, winged.

Figure 4-132 *Firmiana simplex* 梧桐

Reevesia pubescens Mast. 梭罗树（Figure 4-133）

Habit: trees, evergreen.

Bark: gray, exfoliating longitudinally.

Twig: yellow stellate puberulent when young.

Leaf: elliptic-ovate to elliptic, base obtuse or rounded, apex acuminate or acute, pinnate vein, distinctly three-nerved at the base.

Flower: inflorescence a terminal panicle. Flowers creamy white, with densely brownish puberulent. Filaments long, out of corolla.

Fruit: capsule, pyriform or oblong-pyriform, densely brownish puberulent.

Distribution: southwest China, Hainan and Hunan.

Figure 4-133 *Reevesia pubescens* 梭罗树

4.31 Malvaceae 锦葵科

Plants usually with peltate scales or stellate hairs. Leaves simple, usually palmatelyveined, entire or various lobed. Epicalyx often present, 3-to many lobed. Stamens many, filaments connate into tube. Carpels connate or separated, spirally arranged on axis. Fruit a loculicidal capsule or a schizocarp, separating into individual mericarps.

There are 19 genera and 81 species in China.

Hibiscus Linn. 木槿属

Epicalyx lobes 5 to many, free or connate at base, persistent after blooming. Corolla usually large and showy, variously colored. Filament apex truncate or 5-dentate. Style branches 5. Fruit a capsule, valves 5, dehiscence loculicidal. Seeds hairy or glandular verrucose.

Hibiscus syriacus Linn. 木槿（Figure 4-134）

Habit: shrubs, deciduous.

Bark: fairly smooth with brown and gray striping.

Twig: moderate, light gray-brown to brown, raised leaf scar, buds small and not evident.

Leaf: alternate, simple, coarsely serrated and often three-lobed, ovate or diamond shaped, palmately veined from the base, green above, slighter paler below.

Flower: very showy, 5-petaled, ranging from white to reddish-purple depending on cultivar, perfect.

Fruit: capsule, ovoid-globose, densely yellow stellate puberulent.

Distribution: native to middle parts of China, but cultivated all over China.

Figure 4-134　*Hibiscus syriacus* 木槿

Hibiscus mutabilis Linn. 木芙蓉（Figure 4-135）

Habit: shrubs or small trees, deciduous.

Bark: greyish white, smooth.

Twig: densely stellate and woolly pubescent, also in petioles, pedicel, epicalyx, and calyx.

Leaf: broadly ovate to round-ovate, or cordate, 5-7-lobed, papery, abaxially densely stellate minutely tomentose, adaxially sparsely stellate minutely hairy, lobes triangular, basal veins 7-11, margin obtusely serrate, apex acuminate.

Flower: large, solitary. Epicalyx lobes 8, filiform, connate at base. Calyx campanulate. Corolla white or reddish, becoming dark red. Styles 5, pilose.

Fruit: capsule, flattened globose, yellowish hispid and woolly; mericarps 5.

Distribution: native to Fujian, Guangdong, Hunan, Taiwan and Yunnan; cultivated widely in China.

Figure 4-135 *Hibiscus mutabilis* 木芙蓉

4.32 Theaceae 山茶科

Leaves simple, alternate, leathery or so. Flower bracteoles under sepal usually. Stamens many, adnate to petal usually. Fruit a loculicidal capsule or indehiscent and drupaceous or baccate, with 1 to many seeds per locule; pericarp woody, leathery, or fleshy.

There are 12 genera and 274 species in China.

1. *Camellia* Linn. 山茶属

Camellia sinensis (Linn.) O. Ktze. 茶 (Figure 4-136)

Habit: trees, evergreen.

Bark: rough and slightly gray.

Twig: grayish yellow, glabrous; current year branchlet purplish red, white pubescent.

Leaf: elliptic or oblong–elliptic, leathery, abaxially pale green and glabrous or pubescent, adaxially dark green, shiny, and glabrous, base cuneate to broadly cuneate, margin serrate to serrulate, apex bluntly acute to acuminate and with an obtuse tip.

Flower: axillary, solitary or to 3 in a cluster, fragrant. Pedicel short, petal full white. Stamens very numerous, with yellow anthers. Ovary globose, densely white pubescent.

Fruit: capsules, oblate.

Distribution: native to southern China, but cultivated widely in China now.

* The species is similar to *Camellia oleifera* C. Abel 油茶, but the latter leaf blade thick leathery, flowers sessile.

Figure 4-136 *Camellia sinensis* 茶

Camellia japonica Linn. 山茶 (Figure 4-137)

Habit: small trees, evergreen.

Bark: smooth light brown to gray-brown.

Twig: moderately stout, light brown, glabrous; flower buds quite large with imbricate scales, fuzzy greenish brown, vegetative buds much smaller.

Leaf: alternate, simple, evergreen, elliptical to ovate, pointy tip, finely but sharply serrated, leathery, dark shiny green above, green below.

Flower: subsessile. Attractive, with numerous roselike petals, color and shape (single, double, etc.) varies with cultivar, colors ranging from red, pink to white, centers with yellow anthers. Ovary glabrous.

Fruit: dry, round, woody capsule, initially green but ripening in the fall to a light brown, not showy.

Distribution: native to east China, Sichuan.

Figure 4-137 *Camellia japonica* 山茶

Camellia petelotii(Merr.) Sealy 金花茶 II (Figure 4-138)

Habit: shrubs or small trees, evergreen.

Twig: grayish brown; current year branchlets purplish brown, 2-3 mm thick, glabrous.

Leaf: elliptic, oblong-elliptic, or oblong, leathery, both surfaces glabrous, midvein abaxially elevated and adaxially flat or slightly impressed, base broadly cuneate to subrounded, margin serrulate, apex shortly caudate.

Flower: solitary or paired. Sepals 5, leathery, glabrous. Petals 10-14, golden yellow, fleshy. Stamens numerous. Ovary globose, glabrous, 3-loculed; styles 3, distinct.

Fruit: capsule oblate, 3-loculed with 3 seeds per locule, apex sunken; pericarp thick when dry, woody.

Figure 4-138 *Camellia petelotii* 金花茶

2. *Schima* Reinw. ex Blume 木荷属

Trees, evergreen. Anther dorsifixed. Fruit a capsule globose; sepals not persistent in fruit or if persistent not enveloping fruit. Seeds small, reniform, flat, with a marginal membranous wing.

Schima superba Gardn. et Champ. 木荷 (Figure 4-139)

Habit: trees, evergreen.

Bark: dark grey, ruggedly cracked into small, thick, angular pieces.

Twig: glabrous or slightly pubescent.

Leaf: thick leathery, elliptic to oblong, upper surface shiny green, margin undulate, somewhat crenate, apex acuminate.

Flower: inflorescence a raceme. Bracteoles 2, caducous, sepals suborbicular, petals white, obovate, ovary tomentose.

Fruit: capsule, subglobose.

Distribution: southern provinces of China, especially in subtropical area.

Figure 4-139　*Schima superba* 木荷

4.33　Actinidiaceae 猕猴桃科

Trees, shrubs, or woody vines. Leaves alternate, simple, exstipulate. Ovary superior, locules and carpels 3-5 or more; placentation axile; ovules anatropous; styles as many as carpels, distinct or connate, generally persistent. Fruit a berry or leathery capsule.

There are 3 genera and 66 species in China.

Actinidia Lindl. 猕猴桃属

Woody vines, piths solid or lamellate. Winter buds small, enclosed in swollen base of petiole or exposed. Styles as many as carpels (15–30), usually reflexed, persistent, radiating. Fruit a berry.

Actinidia chinensis **Planch.** 中华猕猴桃 II (Figure 4-140)

　　Habit: woody vines, deciduous.

　　Twig: reddish; pith whitish to brown, large, lamellate.

　　Leaf: broadly ovate to broadly obovate or suborbicular, abaxially tomentose, base rounded to truncate, apex truncate to emarginate to abruptly cuspidate.

　　Flower: inflorescence cymose. Flowers orange–yellow. Sepals 5, both surfaces densely yellowish tomentose. Petals 5, shortly clawed at base.

　　Fruit: berry, subglobose to cylindric, densely tomentose.

　　Distribution: south of the Yangtze River.

Figure 4-140　*Actinidia chinensis* 中华猕猴桃

Actinidia eriantha **Benth.** 毛花猕猴桃 (Figure 4-141)

　　Habit: woody vines, deciduous.

　　Twig: densely tomentose with milky–white to dirty yellow hairsin whole plant;

pith white, lamellate.

Leaf: ovate to broadly ovate, papery, densely white stellate tomentose, base rounded or truncate, margin callose-serrulate, apex acute to shortly acuminate.

Flower: inflorescence a cymose. Flower rose-pink usually, with densely lacteous-tomentose.

Fruit: berry, persistently milky-white tomentose; persistent sepals reflexed.

Distribution: south China and east China.

Figure 4-141 *Actinidia eriantha* 毛花猕猴桃

4.34 Dipterocarpaceae 龙脑香科

Gigantic trees with an abundant resin. Flower actionomorphic, hermaphrodite, hypogynous; sepals 5, polysepalous, persistent; petals 5, polypetalous; stamens many in one to several whorls, slightly polyandrous; carpels 3, syncarpous, superior. Fruit usually nutlike, with persistent, variously accrescent calyx.

There are 5 genera and 12 species in China.

Parashorea Kurz 柳安属

Leaf tertiary veins generally scalariform, venation not plicate. Flower sepals with basal thickening appressed to nut. Calyx in fruit without a tube. Ovary without distinct stylopodium. Fruit sepals subequal, imbricate, calyx segments enlarged into wings.

Parashorea chinensis H. Wang 望天树 I (Figure 4-142)

Habit: trees, evergreen, lofty emergents.

Bark: gray or brown, shallowly longitudinally fissured on upper part, but exfoliating in masses on lower part.

Twig: gray to yellowish brown scurfy pubescent or tomentose, lenticels orbicular.

Leaf: elliptic-lanceolate, leathery, both surfaces scurfy pubescent or tomentose, base rounded, margin entire, apex acuminate.

Flower: inflorescence an axillary or terminal cymose panicle, densely grayish yellow scurfy-pubescent or tomentose, flowers sweetly scented. Petals yellowish white; anthers linear-lanceolate. Ovary narrowly ovoid; style columnar, glabrous; stigma small, slightly 3-lobed.

Fruit: nutlike, ellipsoid, densely silvery silky-pubescent; calyx segments subequal, winglike.

Distribution: Guangxi and Yunnan.

Figure 4-142 *Parashorea chinensis* 望天树

4.35　Ericaceae 杜鹃花科

　　Evergreen or deciduous shrubs. Leaves alternate usually, exstipulate. Flowers (4 or) 5-merous. Calyx 5, imbricate, persistent. Corolla 5-lobed, connate. Stamens as many as corolla lobes or double, pollen in tetrads, rarely single. Fruit a capsule or berry, rarely a drupe.

　　There are 22 genera and 826 species in China.

1. *Rhododendron* Linn. 杜鹃属

　　Shrubs or trees. Leaves alternate, sometimes clustered at stem apex; margin entire. Corolla funnelform, campanulate, tubular, rotate or hypocrateriform, regular or slightly zygomorphic, 5-lobed usually, lobes imbricate in bud. Fruit a capsule, dehiscent from top.

Rhododendron ovatum (Lindl.) Planch. ex Maxim. 马银花（Figure 4-143）

　　Habit: shrubs or small arbor, evergreen.

　　Twig: young shoots sparsely glandular-hairy and pubescent.

Figure 4-143　*Rhododendron ovatum* 马银花

Leaf: ovate to oblong-elliptic, margin slightly curved, apex acute and mucronate, adaxial surface glabrous or pubescent.

Flower: calyx deeply 5-lobed; corolla purplish white, with purple flecks inside tube, single in the axil of branchlet apex.

Fruit: capsule, broadly ovoid, densely gray-brown pubescent, also sparsely glandular.

Distribution: south of the Yangtze River and Taiwan.

***Rhododendron simsii* Planch.** 杜鹃 (**Figure 4-144**)

Habit: shrubs, deciduous.

Twig: many and fine, densely shiny brown appressed-setose, flat.

Leaf: ovate to elliptic-ovate, both with coarsely appressed-hairy, margin slightly revolute, finely toothed.

Flower: inflorescence 2-6-flowered. calyx deeply 5-lobed, lobes triangular-long-ovate, coarselyappressed-hairy. Corolla broadly funnelform, rose, bright to dark red, with dark red flecks on upper lobes.

Fruit: capsule, ovoid, calyx persistent.

Figure 4-144　*Rhododendron simsii* 杜鹃

Distribution: each province along the Yangtze River, but also in Taiwan, Sichuan and Yunnan.

* The species is similar to *Rhododendron mariesii* Hemsl. et Wils. 满山红, but the branchlet glabrous and corolla purple for the latter.

2. *Vaccinium* Linn. 越橘属

Inflorescence a terminal or axillary, racemose, fasciculate. Bracts and bracteoles persistent or caducous. Corolla urceolate, campanulate, or tubular. Fruit a several seeded globose berry.

***Vaccinium bracteatum* Thunb.** 南烛(乌饭树) (**Figure 4-145**)

Habit: shrubs or small trees, evergreen.

Twig: many branchlets, pubescent or glabrous when young.

Leaf: elliptic, rhombic- or lanceolate-elliptic, thinly leathery, glabrous, base cuneate, margin plane, denticulate, apex acute.

Flower: inflorescence pseudoterminal, racemose, densely pubescent, many flowered; bracts persistent or caducous, leaflike. Corolla white, densely pubescent. Anthers without spurs.

Fruit: berry, dark purple, pubescent.

Distribution: south of the Yangtze River, but also in Taiwan and Guangdong.

Figure 4-145 *Vaccinium bracteatum* 南烛

Vaccinium mandarinorum Diels 江南越橘（Figure 4-146）

Habit: shrubs or small trees, evergreen.

Twig: terete, glabrous or pubescent, sometimes densely puberulous.

Leaf: ovate or oblong-lanceolate to lanceolate, leathery, both surfaces glabrous, sometimes puberulous on midvein, base cuneate to rounded, margin dentate, apex acute to abruptly acuminate.

Flower: inflorescence a racemose, many flowered. Hypanthium, glabrous. Corolla white or pinkish, tubular or urceolate, glabrous, lobed.

Fruit: berry, dark purple, glabrous.

Distribution: from east China to southwest of China.

Figure 4-146   *Vaccinium mandarinorum* 江南越橘

4.36   Hypericaceae 金丝桃科

Leaves simple, opposite, entire usually, dotted with black or translucent glandular spots sometimes. Stamens many, connate usually. Fruit a capsule, rarely berry.

There are 6 genera and ca. 60 species in China.

Hypericum Linn. 金丝桃属

Invariably opposite or whorled leaves, often dotted with pellucid glands. Stamens numerous, often grouped into three or five bundles, opposite to petals. Fruit a septicidal capsule.

Hypericum monogynum Linn. 金丝桃 (**Figure 4-147**)

Habit: shrubs, deciduous.

Twig: slender, reddish-brown.

Leaf: oblong to elliptic, thickly papery, laminar glands very small dots, abaxial gland absent, main lateral veins 2- or 3-paired, base cuneate to subangustate, apex acute to rounded.

Flower: inflorescence 1-15-flowered. Petals golden yellow to lemon yellow. Stamen fascicle each with 25-35. Styles slender, united nearly to apices then outcurved or very rarely to half free.

Fruit: capsule, broadly ovoid.

Distribution: many areas of China.

Figure 4-147 *Hypericum monogynum* 金丝桃

4.37 Punicaceae 石榴科[①]

Shrubs or small trees, deciduous; branches often terminating as spines. Floral tube campanulate or tubular, leathery; sepals thick and fleshy, persistent. Petals wrinkled, inserted in sepals. Stamens numerous. Ovary inferior. Fruit a berry, with leathery rind. Seeds many, translucent sarcotesta.

There are only 2 species in the world, now widespread in cultivation.

Punica Linn. 石榴属

The morphological characteristics are the same as the family's.

Punica granatum Linn. 石榴 (Figure 4-148)

Habit: shrubs or small trees, deciduous.
Bark: dark gray, or grayish brown, dehiscent.

Figure 4-148 *Punica granatum* 石榴

[①] Punicaceae has been merged into Lythraceae 千屈菜科 in '*Flora of China*', but we still treated it as an independent family in the textbook.

Twig: 4-angled, becoming terete with age, often terminating as indurate spines.

Leaf: petiole short, or sessile; leaf blade adaxially shiny, lanceolate, or oblong, base attenuate, apex obtuse or mucronate.

Flower: campanulate-urceolate. Sepal, erect, orange, thick. Petal multi-colored. Stamens numerous.

Fruit: berry, globose, leathery, crowned by persistent sepals, irregularly dehiscent.

Distribution: cultivated widely in China.

4.38　Aquifoliaceae 冬青科

Trees or shrubs. Leaves simple, alternate, stipules minute. Inflorescence a cyme. Flowers dioecious. Fruit a berrylike drupe, pyrenes 2- numerous.

There are 1 genus and 204 species in China.

Ilex Linn. 冬青属

The morphological characteristics are the same as the family's.

Ilex macrocarpa Oliv. 大果冬青 (Figure 4-149)

Habit: trees, deciduous.

Bark: grayish white or grayish brown, smooth.

Twig: lenticel conspicuous; spur branchlets.

Leaf: ovate to oblong-elliptic, papery, stipule minute, obscure, base rounded or obtuse, margin shallowly serrate, apex acuminate.

Figure 4-149　*Ilex macrocarpa* 大果冬青

Flower: white, small. Male inflorescence, cyme, or solitary, or fasciculate, axillary in the 1st to 2nd year's branchlets; female flower, solitary, axillary on leaves or scales.

Fruit: drupes, black, globose, small; calyx and stigma persistent; peduncle 1–1.5 cm.

Distribution: east China, south China, southwest China, Henan, Hunan and Shaanxi.

Ilex cornuta Lindl. et Paxt. 枸骨 (Figure 4-150)

Habit: shrubs or small trees, evergreen.

Bark: grayish white, smooth.

Twig: young branchlets longitudinally ridged and sulcate; older branchlets subterete. Leaf scars raised, lenticels absent.

Leaf: shiny, quadrangular-oblong, thickly leathery, base rounded or subtruncate, margin with 1 or 2 spines per side, apex with 1 strong spine reflexed usually.

Flower: small, yellowish green; inflorescence a cyme, fasciculate, axillary in the 2nd year's branchlets.

Fruit: drupe, red, globose, calyx and stigma persistent.

Distribution: from central China to east China.

Figure 4-150 *Ilex cornuta* 枸骨

Ilex chinensis Sims 冬青（Figure 4-151）

Habit：trees, evergreen.

Bark：grayish black or pale gray.

Twig：terete, thinly angular glabrous; lenticels obscure, small.

Leaf：elliptic to lanceolate, thinly leathery, glabrous, petiole dark purple usually; lateral veins 6-9 pairs, base cuneate, margin crenate, apex acuminate.

Flower：small, purplish or purple-red, petals reflexed at anthesis. Inflorescences a cyme, or solitary, axillary on current year's branchlets, glabrous.

Fruit：drupes red, narrowly globose.

Distribution：south of the Yangtze River-Huaihe River.

Figure 4-151　*Ilex chinensis* 冬青

Ilex latifolia Thunb. 大叶冬青（Figure 4-152）

Habit：trees, evergreen, glabrous throughout.

Bark：bark gray-black, smooth.

Twig：branchlet sturdy, longitudinally ridged and sulcate; leaf scars conspicuous.

Leaf: oblong to ovate-oblong, thickly leathery, midvein impressed adaxially and obvious adaxially, margin sparsely serrate, stipules very minute.

Flower: inflorescence a cyme, axillary in the 2nd year's branchlet. Petals ovate, united in base.

Fruit: drupe red or brown, globose, small; stigma and calyx persistent.

Distribution: south provinces of the Yangtze River and Henan.

Figure 4-152 *Ilex latifolia* 大叶冬青

4.39 Celastraceae 卫矛科

Leaves simple, alternate or opposite. Inflorescence an axillary or terminal, cymose, thyrsoid, or racemose. Floral disk conspicuous, diversiform. Fruit a capsule, drupe, berry, or samara. Aril conspicuous, enveloping seed usually.

There are 14 genera and 192 species in China.

Euonymus Linn. 卫矛属

Leaves opposite usually. Flowers 4-5-merous, filaments short, disk fleshy, stigma 3-5-lobed. Inflorescence a cyme, axillary. Capsule valvular dehiscence, prickly sometimes, laterally winged; aril basal to enveloping seed.

Euonymus alatus (Thunb.) Sieb. 卫矛 (Figure 4-153)

Habit: shrubs, deciduous.

Twig: 4-angled; young branches usually with 2 or 4 winglike corks.

Leaf: obovate to obovate-elliptic, simple, opposite, thinly leathery, petiole sessile or very short, base cuneate, margin crenulate, apex acute.

Flower: small, yellowish green. Inflorescence a cyme, with 3 flowers usually.

Fruit: capsule, reddish brown when fresh, 4-lobed deeply. Seeds brown with red aril.

Distribution: many areas of China.

Figure 4-153 *Euonymus alatus* 卫矛

Euonymus maackii **Rupr.** 白杜(丝棉木)(**Figure 4-154**)

Habit: trees, deciduous.

Bark: gray, grayish brown.

Twig: terete, sturdy, green to light green.

Leaf: ovate to elliptic -lanceolate, thinly leathery, base subattenuate, margin crenulate, apex acuminate.

Flower: small, yellowish green, anthers purplish. Inflorescence a cymose, 1-2 dichotomously branched, 3-7 flowered.

Fruit: capsule, rhombic, pinkish, 4-angled. Seeds with orange aril.

Distribution: from Heilongjiang to north China, central China and east China.

Figure 4-154 *Euonymus maackii* 白杜

Euonymus japonicus **Thunb.** 冬青卫矛(大叶黄杨) (**Figure 4-155**)

Habit: shrubs to trees, evergreen.

Twig: green to light green, glabrous, 4-angled nearly.

Leaf: ovate to obovate, leathery, shiny, base orbicular, margin crenulate distally, apex orbicular, lateral veins 6-8 pairs, slightly visible or unclear.

Flower: small, greenish white. Inflorescence a cymose with 5-12 flowers, axillary.

Fruit: capsule, globose. Seeds dark brown, aril orange-red.

Distribution: many areas of China.

Figure 4-155   *Euonymus japonicus* 冬青卫矛

Euonymus fortunei **(Turcz.) Hand.-Mazz.** 扶芳藤 (**Figure 4-156**)

Habit: subshrubs, evergreen, ascending or procumbent usually.

Twig: brown or green-brown, sometimes striate.

Leaf: variously ovate; petiole short, or sessile; base truncate, margin crenulate to serrate, apex obtuse, lateral veins invisible.

Flower: inflorescence a cyme, 5-15 flowers in axillae; flowers small, greenish white, 4-merous.

Fruit: capsule, brown to red-brown. Seeds with orange-red aril.

Distribution: from east part to southwest part of China.

Figure 4-156 *Euonymus fortunei* 扶芳藤

4.40 Elaeagnaceae 胡颓子科

Most parts with distinctive silvery or brownish peltate scales and/or stellate hairs, sometimes branches spine-tipped. Petals absent. Fruit a drupelike, with a single seed.

There are 2 genera and 74 species in China.

Elaeagnus Linn. 胡颓子属

Plants sometimes spiny. Flowers clustered on short axillary shoots, sometimes solitary. Calyx tubular, 4-lobed. Fruit a drupe.

Elaeagnus pungens Thunb. 胡颓子 (Figure 4-157)

Habit: shrubs, evergreen.

Bark: grayish.

Twig: densely branched with brown scaly; thorns usually.

Leaf: oblong, leathery, petiole with brown scaly, abaxially with dense whitish and usually also brown scales, base rounded, undulate margins, apex obtuse.

Flower: flowers 1–4 clustered in axils; pedicel short, brown scaly; calyx tube funnel form, lobes ovate.

Fruit: drupe, oblong, brown scaly.

Distribution: south of the Yangtze River.

* Sometimes, we also meet *Elaeagnus argyi* H. Lévl. 佘山羊奶子, *Elaeagnus umbellate* Thunb. 牛奶子, and *Elaeagnus multiflora* Thunb. 木半夏 in the field, but they are all deciduous.

Figure 4-157　*Elaeagnus pungens* 胡颓子

4.41　Rhamnaceae 鼠李科

Trees or shrubs, thorny often. Stipules small, caducous or persistent, sometimes transformed into spines. Calyx tube patelliform, 5-lobes on top, rarely 4-lobes. Ovary superior, 2-4-loculed. Fruit a drupe, berry, nut or capsule.

There are 13 genera and 137 species in China.

1. *Hovenia* Thunb. 枳椇属

Deciduous trees, without spine. Long petiolate, 3-veined from leaves base. Inflorescence axis twisty and fleshy, when fruit. Fruit a drupe.

***Hovenia acerba* Lindl.** 枳椇(拐枣)(**Figure 4-158**)

Habit: trees, deciduous, tall.

Bark: dark gray.

Twig: brown or black-purple, conspicuous white lenticels.

Leaf: broadly ovate or other, papery, petiole long, margin finely serrulate.

Flower: symmetrical, dichasial cymose panicles; flowers small, yellowish green; style deeply branched.

Fruit: brown at maturity, fruiting peduncles and pedicels dilated and fleshy.

Distribution: from east China to southwest China.

Figure 4-158 *Hovenia acerba* 枳椇

2. *Paliurus* Mill. 马甲子属

3-veined from base, stipules usually changed into 1 or 2 lignified. Fruit a drupe, woody, with wing around the drupe.

Paliurus hemsleyanus Rehd. ex Schir. et Olabi 铜钱树（**Figure 4-159**）

Habit：trees or shrubs, deciduous.
Bark：dark gray.
Twig：glabrous, spine or not.
Leaf：broadly ovate to broadly elliptic, alternate, stipular spines in vegetative shoots, 3-veined from base, base oblique often, margin crenate, apex acuminate to acute.
Flower：small, yellowish green, glabrous, in cymes or cymose panicles, terminal or axillary.
Fruit：drupe, disk-shaped, wing thinly papery to leathery
Distribution：east China and south China.

Figure 4-159 *Paliurus hemsleyanus* 铜钱树

3. *Sageretia* Brongn. 雀梅藤属

Shrubs scandent, unarmed or spinescent. Simple leaves, simple, alternate or subopposite. Flowers small, sessile or subsessile; inflorescence panicle or spica.

Sageretia thea (Osbeck) M. C. Johnst. 雀梅藤 (Figure 4-160)

    Habit: shrubs, scandent.

    Bark: dark gray.

    Twig: spine, brownish, slender, finely tomentose when young.

    Leaf: small, elliptic, oblong, or ovate-elliptic, subopposite, petiole short, leaf blade papery, apex acute, base rounded, margin finely serrulate.

    Flower: greenish white, small, sessile.

    Fruit: drupe, black or purple-black at maturity.

    Distribution: from east China to southwest China.

Figure 4-160 *Sageretia thea* 雀梅藤

4. *Rhamnus* Linn. 鼠李属

Often spinose. Flowers small, bisexual or unisexual, solitary or few fascicled, or cyme. Fruit a berrylike drupe with 2-4 stoned; seed furrowed.

***Rhamnus utilis* Decne.** 冻绿 (**Figure 4-161**)

Habit: small tree, deciduous.

Bark: gray.

Twig: terminating in a spine often; old branches brown or purple-red.

Leaf: elliptic or oblong, papery, opposite to subopposite or fascicled on short shoots, margin finely serrate, apex acute, lateral veins 5-8 pairs.

Flower: yellowish green, unisexual, 4-merous, fascicled usually.

Fruit: drupe, black at maturity; seeds with furrow at base.

Distribution: east China, south China and southwest China.

Figure 4-161 *Rhamnus utilis* 冻绿

4.42 Vitaceae 葡萄科

Woody climbers or herbaceous climbers, rarely shrubs. Tendril simple, bifurcate to trifurcate, usually leaf-opposed. Inflorescence a panicle or corymb, often leaf-opposite. Fruit a berry.

There are 8 genera and 146 species in China.

1. *Vitis* Linn. 葡萄属

Woody lianas; bark lenticels inconspicuous, pith brown. Tendril leaf-opposed, usually bifurcate. Inflorescence a thyrse. Petals 5, united at apex and shed as a cap at anthesis. Fruit a berry.

***Vitis vinifera* Linn.** 葡萄 (**Figure 4-162**)

Habit: lianas, deciduous.

Bark: gray or brown, sometimes shredding.

Twig: longitudinal ridges, tendrils bifurcate.

Leaf: oval, simple, basal veins 5, base deeply cordate, margin with rough serrates, large, irregular, apex acute.

Flower: panicle dense, pendulous; flowers polygamous or dioecious.

Fruit: berry.

Figure 4-162 *Vitis vinifera* 葡萄

Distribution: native to West Asia, cultivated all over China.

* *Vitis davidii* (Romanet du Caillaud) Föex 刺葡萄 and *Vitis romanetii* Romanet du Caillaud 秋葡萄 are usually found in the field; the former conspicuously prick or tuberculate in branchlets, and the latter glandular hairs apparently.

2. *Ampelopsis* Mich. 蛇葡萄属

Woody lianas. Tendrils inconspicuous inflated; pith white. Flowers green, small; inflorescence a corymbose cyme, leaf-opposed or pseudoterminal. Fruit a berry.

Ampelopsis humulifolia Bunge 葎叶蛇葡萄 (**Figure 4-163**)

Habit: lianas, deciduous.

Twig: brown, glabrous, with longitudinal ridges.

Leaf: simple, adaxially bright green, abaxially pale green, 3-5-lobed, cordate; margin with large, sharp teeth, apex acuminate.

Flower: small; inflorescence a cyme, leaf-opposed.

Fruit: berry globose, small.

Distribution: northeast China, north China and east China.

Figure 4-163 *Ampelopsis humulifolia* 葎叶蛇葡萄

4.43 Ebenaceae 柿树科

Leaves simple, alternate, entire. Flower unisexual usually, actinomorphic. Corolla united, 3-7-lobed slightly. Female calyx 3-7-lobed, persistent, enlarged when mature. Ovary superior. Fruit a berry.

There are 1 genera and 60 species in China.

Diospyros Linn. 柿属

Terminal buds absent, unarmed or spine. Berry usually with an enlarged persistent calyx.

***Diospyros rhombifolia* Hemsl.** 老鸦柿 (Figure 4-164)

Habit: tree, deciduous.

Bark: brown, glossy.

Twig: young branchlets pale purple, pubescent; spines.

Leaf: Petiole short, papery, base cuneate, margin ciliate, apex acute to acuminate, reticulate veinlets raised on both surfaces.

Figure 4-164 *Diospyros rhombifolia* 老鸦柿

Flower: white; calyx persistent, oblong-lanceolate when fruited, with conspicuous ridges, reflexed.

Fruit: berry, orange, glabrous, shiny.

Distribution: east China.

Diospyros kaki Thunb. 柿 (Figure 4-165)

Habit: trees, deciduous.

Bark: dark brown, scaly.

Twig: densely pubescent when young.

Leaf: elliptic to obovate, adaxially dark green, abaxially pale green, brown pubescence sparse, base cuneate, apex acuminate, reticulate veinlets clearly defined.

Flower: yellow, calyx as long as corolla; male flowers clustered, female flowers solitary.

Fruit: berry, yellow to orange, globose, glabrescent.

Distribution: native to China.

* *Diospyros kaki* var. *silvestris* Makino 野柿 has densely brown pubescent in young branchlet and petiole, and leaf blade is smaller and thinner than *D. kaki*.

Figure 4-165 *Diospyros kaki* 柿

4.44 Simaroubaceae 苦木科

Bark bitter. Leaves pinnate. Flowers unisexual or polygamous; filaments free, base often with an appendage. Fruit a drupe or samara.

There are 3 genera and 10 species in China.

Ailanthus Desf. 臭椿属

Deciduous. Bud with 2-4 scales, round. Leaves alternate, pinnate. Terminating panicle, large. Fruit a samara.

Ailanthus altissima (Mill.) Swingle 臭椿 (Figure 4-166)

Habit: trees, deciduous.

Bark: smooth and straightly grained.

Twig: reddish brown when young.

Leaf: odd-pinnate, large. Leaflets opposite, blades ovate-lanceolate, abaxially dark green, adaxially gray-green, margin with 1-2 large serrates in base, smelly when rubbed.

Flower: inflorescence a panicle; flowers light green, small.

Fruit: samara oblong.

Distribution: from north China to southeast China.

Figure 4-166 *Ailanthus altissima* 臭椿

4.45 Meliaceae 楝科

Leaves usually pinnate. Inflorescence a panicle. Flower bisexual, rarely unisexual; stamens as 2 times as petals, anthers usually sessile on stamen tube. Fruit a capsule, berry or rarely drupe. There are 17 genera and 40 species in China.

Melia Linn. 楝属

Leaves in spirals, 2- or 3-pinnate. Petals 5-6; staminal tube margin 10-12-lobed; anthers 10-12, inserted between filament tube lobes. 2 superposed ovules per locule. Fruit a drupe.

Melia azedarach Linn. 楝 (Figure 4-167)

Habit: trees, deciduous.
Bark: brownish gray, longitudinally exfoliating.
Twig: stellate hair when young, but disappeared soon.
Leaf: odd-pinnate, 2-pinnate or 3-pinnate, leaflets opposite, ovate to elliptic, margin crenate or sometimes entire.
Flower: inflorescence a panicle. Calyx 5-lobed; petals 5, pale purple; anthers 10.
Fruit: drupe, pale yellow.

Figure 4-167 *Melia azedarach* 楝

Distribution: south of the Yellow River in China.

* *Melia toosendan* Sieb. et Zucc. 川楝 was considered as the other species in southwest China because of the bigger fruit, but it is merged into *M. azedarach* in FOC.

4.46 Sapindaceae 无患子科

Leaves alternate, estipulate, pinnate or palmate, rarely simple. Stamens 8 usually, filaments free or connate, but always present in one side. Ovary superior, 3-loculed usually; ovules 1 or 2 per locule. Fruit a loculicidal capsule, berry, or drupe, or samara.

There are 21 genera and 52 species in China.

1. *Sapindus* Linn. 无患子属

Leaves paripinnate. Flowers small, complete. Fruit a drupe, ellipsoid, parted into 3 schizocarps, pericarp fleshy.

***Sapindus saponaria* Linn.** 无患子 (**Figure 4-168**)

Habit: trees, deciduous.

Bark: grayish brown or blackish brown.

Twig: green, glabrous, with dense lenticels.

Figure 4-168 *Sapindus saponaria* 无患子

Leaf: paripinnate, large. Leaflets 8-12 pairs, papery, elliptic-lanceolate or slightly falcate, base cuneate, slightly asymmetrical, apex acute.

Flower: inflorescence a terminal, conical. Flowers small, calyx 5, petal 5, scales 2, at base adaxially.

Fruit: schizocarp, orange, turn black when dry.

Distribution: south of the Yangtze River.

2. *Koelreuteria* Laxm. 栾树属

Leaves imparipinnate or bipinnate. Thyrse terminal. Capsule swollen, carpel membranous, loculicidal into 3 schizocarps. Seed black, globose, arillodes absent.

***Koelreuteria paniculata* Laxm.** 栾树

Habit: trees, deciduous.

Bark: grayish brown to black; lenticels small, black.

Twig: lenticels conspicuous, branchlets pale brown when young.

Leaf: odd-pinnate. Leaflets 7-15 pairs, opposite or subopposite, ovate to lanceolate, papery, base obtuse to subtruncate, apex acute or shortly acuminate, margin irregularly obtusely serrate, or pinnately lobed.

Flower: inflorescence thyrse, large, flowers small, pale yellow; petals 4 with claw; stamens 8, anthers with sparse pubescence.

Figure 4-169 *Koelreuteria bipinnata* 复羽叶栾树

Fruit: capsule, conical, carpel ovoid, abaxially reticulate veined.

Distribution: widely cultivated in China.

* *Koelreuteria bipinnata* Franch. 复羽叶栾树 (Figure 4-169) is similar to this species, but the former is bipinnate with several serrulate or entire in leaflet margin.

4.47 Anacardiaceae 漆树科

Resiniferous secretory ducts in bark and foliage. Leaves alternate, exstipulate, odd-pinnate usually. Inflorescence a terminal or axillary thyrsoid or panicle. Fruit a drupe.

There are 17 genera and 55 species in China.

1. *Pistacia* Linn. 黄连木属

Deciduous and dioecious. Leaflets entire. Inflorescence a paniculate. Drupe red at maturity, pointed, oblique ovate; endocarp bony.

***Pistacia chinensis* Bunge 黄连木 (Figure 4-170)**

Habit: trees, deciduous.

Bark: grayish brown, peels to reveal salmon inner bark.

Figure 4-170 *Pistacia chinensis* 黄连木

Twig: grayish brown, pubescent, lenticels minute; winter buds red, fragrant.

Leaf: even-pinnate, 10-14 opposite leaflets; leaflet blade lanceolate, papery, base oblique, margin entire, apex acuminate, minutely pubescent along both midribs.

Flower: small, dioecious. Inflorescence a racemose and paniculate in male and female, respectively.

Fruit: drupe, obovate to globose, slightly compressed.

Distribution: north China, northwest China and south of the Yangtze River.

2. *Rhus* Linn. 盐肤木属

Shrubs or trees, Deciduous with latex or watery juice. Leaves imparipinnately compound. Inflorescence a terminal, paniculate or thyrsoid. Flower functionally unisexual or bisexual, 5-merous. Fruit a drupe, slightly compressed, mixed glandular pubescent and pilose, red at maturity.

Rhus chinensis Mill. 盐肤木 (**Figure 4-171**)

Habit: shrubs to trees, deciduous.

Bark: gray or grayish brown.

Twig: dense lenticels, ferruginous pubescent; leaf scars triangled.

Figure 4-171 *Rhus chinensis* 盐肤木

Leaf: odd-pinnate, rachis winged to wingless, ferruginous pubescent; leaflets 7-13, ovate to oblong, base rounded, margin dentate, apex acute.

Flower: small, white. Inflorescence a panicle with multi-branched, densely ferruginous pubescent.

Fruit: drupe, globose, mixed pilose and glandular-pubescent.

Distribution: all over China, excluding Inner Mongolia, Xinjiang and north part of northeast China.

3. *Toxicodendron* Mill. 漆属

White latex in phloem. Inflorescence a paniculate or racemose, axillary, often pendulous at fructification. Drupe subglobose or oblique.

Toxicodendron succedaneum (Linn.) Kuntze 野漆树 (Figure 4-172)

Habit: shrubs to trees, deciduous.

Bark: grayish green, dense lenticels.

Twig: sturdy, glabrous; terminal buds large.

Leaf: odd-pinnate; leaflets blade thinly leathery, oblong-elliptic to ovate-lanceolate, base oblique, margin entire, apex acuminate.

Figure 4-172 *Toxicodendron succedaneum* 野漆树

Flower: small, yellowish green; inflorescences a paniculate, multi-branched, glabrous.

Fruit: drupe, large, asymmetrical, compressed, apex eccentric.

Distribution: south parts of Yellow River.

* This species is similar to *Toxicodendron vernicifluum* (Stokes) F. A. Barkl. 漆树, but there is obvious pubescent on twigs, terminal buds, rachis, petioles and inflorescence in latter. *T. vernicifluum* is a main source of varnish used in the manufacture of lacquer-ware.

4.48 Aceraceae 槭树科

Leaves simple, opposite, palmately lobed or compound leaf. Fruit a double samaras usually. There are 2 genera and 101 species in China.

Acer Linn. 槭属

The morphological characteristics are the same as the family's.

Acer palmatum Thunb. 鸡爪槭 (Figure 4-173)

Habit: trees, deciduous.

Bark: greenish gray or light brown, smooth.

Figure 4-173 *Acer palmatum* 鸡爪槭

Twig: grayish green, glabrous; winter buds purplish red, terminal buds usually absent.

Leaf: simple, leaf blade suborbicular, papery, base cordate to subtruncate, palmately lobed to middle usually, margin irregularly doubly serrate, apex acuminate.

Flower: small, purplish red; ovary glabrous. Inflorescence a corymbose-paniculate.

Fruit: double samaras, wings spreading at obtuse angle.

Distribution: cultivated in China widely.

* This species is a famous small ornamental tree with many formas and varieties, such as *Acer palmatum* f. *atropurpureum* (Van Houtte) Schwerim 红枫, *Acer palmatum* var. *dissectum* (Thunb.) K. Koch 羽毛槭.

Acer buergerianum Miq. 三角槭 (Figure 4-174)

Habit: trees, deciduous.

Bark: dark gray, dehiscent.

Twig: slender, with dense lenticel; winter buds brown, small.

Leaf: simple, glabrous, leaf blade abaxially whitish, 3-lobed, ovate to obovate, papery, base rounded or cuneate, margin entire usually, apex acute, primary veins 3.

Flower: small, yellowish white; ovary densely yellowish villous. Inflorescence a terminal corymbose, pubescent.

Figure 4-174 *Acer buergerianum* 三角槭

Fruit: double samaras, yellowish brown, wing broad at middle, contracted at base.

Distribution: south of the Yangtze River, cultivated in north China and south part of northwest China.

Acer henryi Pax 建始槭(三叶槭) (**Figure 4-175**)

Habit: trees, deciduous.

Bark: gray.

Twig: slender, green or slightly purplish.

Leaf: 3-foliolate; leaflets elliptic or oblong-elliptic, papery, margin entire or remotely serrate apically, apex acuminate, pinnate vein.

Flower: small, yellowish green, andromonoecious; inflorescence a racemose, pendulous, with male and female flowers.

Fruit: double samaras, yellowish brown at maturity; nutlets convex, wings spreading at 90° or erectly.

Distribution: along the Yangtze River basin, Henan, Gansu and Shaanxi.

Figure 4-175　*Acer henryi* 建始槭

Acer truncatum Bunge 元宝槭(**Figure 4-176**)

Habit: trees, deciduous.

Bark: grayish brown or dark brown.

Twig: slender, glabrous; winter buds purplish brown.

Leaf: simple, 5-lobed, glabrous, abaxially pale green, adaxially dark green, papery, base usually truncate, margin entire, apex acuminate, palmate veins.

Flower: small, yellowish green; sepals and petals 5, respectively; inflorescence a corymbose.

Fruit: double samaras, nutlets flat, thick, glabrous; wing usually as long as nutlets, parallel on both sides, wings spreading at obtuse or right angles.

Distribution: northeast China, north China and east China.

* This species is similar to *Acer pictum* subsp. *mono* (Maxim.) H. Ohashi 色木槭, but the wings spreading of latter are more obvious.

Figure 4-176 *Acer truncatum* 元宝槭

4.49 Hippocastanaceae 七叶树科

Leaves opposite, long petiolate, exstipulate; leaf blade palmately 5 – 11 – foliolate; Inflorescence a terminal thyrse, usually erect. Fruit a capsule.

There are 2 genera and 5 species in China.

Aesculus Linn. 七叶树属

The morphological characteristics are the same as the family's.

Aesculus chinensis Bunge 七叶树 (**Figure 4-177**)

Habit: trees, deciduous.

Bark: grayish brown.

Twig: glabrous, but pubescent when young.

Leaf: petiole long, 5-7-foliolate; leaflets sessile, base cuneate, margin shallowly crenate, apex abruptly acuminate, lateral veins many pairs.

Flower: calyx 5; petals 4, unequal, white or pale yellow, with red or yellow spots on top. Inflorescence a thyrse conic.

Fruit: capsule, broadly obovoid or pyriform, brownish yellow, with dense verrucose.

Distribution: south provinces of the Yangtze River and Gansu, cultivated in China widely.

* *Aesculus chinensis* Bunge 七叶树 has been revised as this species.

Figure 4-177 *Aesculus chinensis* 七叶树

4.50　Oleaceae 木犀科

Leaves opposite, rarely alternate, simple, trifoliolate, or pinnately. Calyx 4 usually, rarely 5-15 lobed. Corolla 4 usually, rarely 6-12 lobed. Stamens 2 usually, rarely 3-5, inserted on corolla tube. Fruit a drupe, berry, capsule, or samara.

There are 10 genera and 160 species in China.

1. *Forsythia* Vahl 连翘属

Branches hollow or with lamellate pith; Twigs: with several-angled. Leaves opposite, simple. Flowers 1 to several fascicled in leaf axils. Corolla yellow, 4-parted; tube campanulate. Fruit a capsule.

***Forsythia suspensa* (Thunb.) Vahl 连翘 (Figure 4-178)**

Habit: shrubs, deciduous.

Twig: yellow-brown or gray-brown; internodes hollow; pendulous.

Leaf: simple, ovate to elliptic-ovate, leathery, glabrous or sometimes pubescent, base rounded to cuneate, margin serrate, apex acute.

Figure 4-178　*Forsythia suspensa* 连翘

Flower: solitary or 2 to several in leaf axils. Calyx lobes oblong; corolla yellow; tube subequal to calyx lobes.

Fruits: capsules ovoid to long ellipsoid.

Distribution: north China, east China and northeast China.

* This species is similar to *Forsythia viridissima* Lindl. 金钟花, but the pith of latter is lamellate.

2. *Jasminum* Linn. 素馨属

Leaves simple, or 3 - foliolate, or odd - pinnate. Inflorescence a basically cymose, or corymb. Calyx campanulate, 4 - 9 - lobed; corolla funnelform, 4 - 9 - lobed. Stamens 2. Fruit a berry, didymous.

Jasminum sambac (Linn.) Ait. 茉莉花 (Figure 4-179)

Habit: shrubs, evergreen, erect or scandent.

Twig: terete or slightly compressed, pubescent when young.

Leaf: simple, opposite, orbicular to elliptic, margin entire, papery, both ends blunt; primary veins 4-6 on each side of midrib.

Flower: fragrant; calyx 8 - 9 - lobed, linear; corolla white, tube; bracts subulate. Inflorescence a terminal cyme.

Fruit: berry, purple-black, globose.

Distribution: native to India, but widely cultivated in southern China.

Figure 4-179 *Jasminum sambac* 茉莉花

***Jasminum nudiflorum* Lindl.** 迎春花

Habit: shrubs, deciduous, creeping or forming an intricate cushion.

Twig: 4-angled, somewhat narrowly winged, glabrous.

Leaf: opposite, 3-foliolate, petiole conspicuous, leaflet blade ovate to elliptic, base cuneate, apex acute or obtuse, terminal one sessile or basally decurrent into a short petiolule.

Flower: solitary, axillary or rarely terminal. Calyx green, lobes 5 or 6; corolla yellow, tube, lobes 5 or 6, oblong or elliptic.

Fruit: berry, ovoid or ellipsoid.

Distribution: native to southwest China, but cultivated in many areas of China.

* This species is similar to *Jasminum mesnyi* Hance 野迎春(云南黄素馨) (Figure 4-180), but the latter is evergreen, flower larger and corolla more obvious.

Figure 4-180 *Jasminum mesnyi* 野迎春

3. *Osmanthus* Lour. 木犀属

Evergreen. Inflorescence a cymose, fascicled in leaf axils or in very short and axillary or terminal panicle. Corolla tube short, 4-lobed, imbricate in bud. Fruit a drupe.

Osmanthus fragrans Loureiro 木犀(桂花) (Figure 4-181)

Habit: trees or shrubs, evergreen.

Bark: grayish brown.

Twig: yellowish brown, glabrous.

Leaf: elliptic to elliptic-lanceolate, base cuneate, margin entire or usually serrulate along distal half, apex acuminate; 6-10 primary veins adaxially impressed and abaxially raised.

Flower: inflorescence a cyme fascicled in leaf axils, many flowered. Calyx obscure; corolla pale yellow; tube short.

Fruit: drupe, purple-black, ellipsoid.

Distribution: native to southwest China and central China, widely cultivated in many areas of China.

* There are many cultivars, including four groups: *Osmanthus Siji* Group, *Osmanthus Albus* Group, *Osmanthus Leteus* Group, and *Osmanthus Auranticus* Group.

Figure 4-181 *Osmanthus fragrans* 木犀

4. *Ligustrum* Linn. 女贞属

Leaves simple, opposite, entire. Inflorescence a terminal panicle of cyme. Fruit a berrylike drupe with membranous or papery endocarp.

Ligustrum lucidum **W. T. Ait.** 女贞 (Figure 4-182)

　　Habit: trees, evergreen.

　　Bark: yellowish brown, gray or purplish red.

　　Twig: terete, with sparse lenticels.

　　Leaf: ovate to elliptic-lanceolate, leathery, base rounded, apex acute, primary veins slightly raised or obscure, margin entire.

　　Flower: white, sessile; calyx small, corolla tube as long as calyx. Inflorescence a panicle terminal.

　　Fruit: berry, deeply blue-black.

　　Distribution: south provinces of the Yangtze River.

Figure 4-182 *Ligustrum lucidum* 女贞

4.51　Apocynaceae 夹竹桃科

　　Woody plants with latex or rarely watery juice. Leaves simple, opposite, rarely whorled or alternate, pinnately veined. Corolla 5-lobed, rotating arrangement usually, lobes overlapping to right or left. Pollen granular. Fruit a berry, drupe, capsule, or follicle.

　　There are 44 genera and 145 species (38 endemic) in China. Most of them grow in the south and southwest of China.

1. *Nerium* Linn. 夹竹桃属

　　Shrubs evergreen, juice watery. Leaves in whorls of 3 or opposite, lateral veins numerous, pinnate parallel. Corolla funnelform; corona segments 5, petal-like, large, fringed. Stamens inserted at apex of corolla tube; anthers with a bristly, filiform apical appendage. Fruit a follicle.

***Nerium oleander* Linn.** 欧洲夹竹桃 (**Figure 4-183**)

　　Habit: shrubs, evergreen.
　　Twig: grayish green and juice watery.

Figure 4-183　*Nerium oleander* 欧洲夹竹桃

Leaf: very narrowly elliptic, leathery, base cuneate or decurrent on petiole, apex acuminate or acute.

Flower: fragrant. Sepal narrowly triangular to narrowly ovate; corolla purplish red, pink, white, salmon, or yellow; lobes single or double.

Fruit: follicle, cylindric.

Distribution: native to Iran, Nepal and India; widely cultivated in south provinces of China.

* There is extremely toxic in all parts of the plant. *Nerium indicum* Mill. 夹竹桃 has been revised as this species.

2. *Trachelospermum* Lem. 络石属

Lianas woody, latex white. Leaves opposite. Calyx small, deeply divided, basal glands 5-10; corolla salverform, without corona in throat. Follicles 2, linear or fusiform. Seed terminal with hair.

Trachelospermum jasminoides (Lindl.) Lem. 络石 (Figure 4-184)

Habit: lianas woody, evergreen.

Stem: brownish, lenticellate; young twig pubescent, glabrous when older.

Leaf: ovate to obovate or narrowly elliptic, papery.

Flower: sepal spreading or reflexed; corolla white, tube dilated at middle; stamens inserted at middle of corolla tube. Inflorescence a cyme paniculate, terminal or axillary.

Fruit: follicle, linear.

Distribution: south of the Qinling Mountains-Huaihe River.

Figure 4-184 *Trachelospermum jasminoides* 络石

4.52 Rubiaceae 茜草科

Leaves simple, opposite or in whorls. Stipules present, usually interpetiolar, sometimes united around stem into a sheath. Flower usually bisexual, actinomorphic; calyx gamosepalous, 4-5-lobed; corolla gamopetalous, often with a long tube, 4-5-lobed. Fruit a capsule, berry or drupe.

There are 97 genera and 701 species in China. Most of them grow from southeast China to southwest China.

1. *Gardenia* J. Ellis 栀子属

Leaves opposite or in whorl; stipules persistent or deciduous, united shortly around stem or united completely into a conical cap. Ovary superior, 1-celled. Fruit leathery or fleshy, with longitudinal ridges and persistent calyx limb usually.

Gardenia jasminoides **J. Ellis 栀子（Figure 4-185）**

Habit: shrubs, evergreen.

Twig: gray to grayish white, terete to flattened, with internodes developed to shortened.

Figure 4-185 *Gardenia jasminoides* 栀子

Leaf: opposite or rarely ternate, thinly leathery, oblong-lanceolate to elliptic, base cuneate, apex acute to long acuminate, margin entire, stipules calyptrate, cylindrical.

Flower: solitary, terminal. Corolla white to pale yellow, simple; tube cylindrical, in throat pilose; lobes 5-numerous, obovate or obovate-oblong.

Fruit: berry, yellow or orange-yellow, with 5-9 longitudinal ridges and persistent calyx lobe.

Distribution: from south of the Yangtze River to southwest China.

* *Gardenia stenophylla* Merr. 狭叶栀子 is planted in many cities of China, its individuals small, leaf blade narrowly lanceolate to linear-lanceolate, and fruit weak longitudinal ridges.

2. *Emmenopterys* Oliv. 香果树属

Some flowers of most inflorescences with 1 lobe expanded into a white petaloid calycophyll. Seeds numerous, flattened, winged.

Emmenopterys henryi Oliv. 香果树 Ⅱ (Figure 4-186)

Habit: trees, deciduous.

Bark: grayish brown.

Twig: angled to terete, often lenticellate, stout, glabrous.

Leaf: drying papery or leathery, broadly elliptic to ovate-oblong, base acute to cuneate, apex acute, secondary veins 5-9 pairs, in abaxial axils with pilosulous domatia.

Flower: bract caducous, narrowly triangular, acute. Calyx lobate sometimes, white, conspicuous and persistent when mature.

Figure 4-186 *Emmenopterys henryi* 香果树

Fruit: Capsule, oblong, longitudinally weakly ribbed; seeds numerous, broadly winged.
Distribution: south of the Qinling Mountains.

4.53 Bignoniaceae 紫葳科

Corolla 5-lobed, oblique. Stamens as same as lobes, opposite, usually inserted in the corollatube. Capsule dehiscing. Seeds flat, usually winged or with tufts of hairs at both ends.

There are 12 genera and 35 species in China.

Catalpa Scop. 梓属

Trees deciduous. Leaves simple, opposite or 3-whorl, purple glandular punctate at vein axils abaxially. Calyx bilabiate or irregularly divided; corolla campanulate, bilabiate, upper lip 2-lobed, lower lip 3-lobed. Fertile stamens 2, inserted at base of corolla tube. Fruit a capsule, dehiscent. Seeds with hair.

Catalpa ovata G. Don 梓树 (Figure 4-187)

Habit: trees, deciduous.

Bark: grayish brown, longitudinally dehiscent.

Figure 4-187 *Catalpa ovata* 梓树

Twig: sparsely pubescent when young.

Leaf: opposite or nearly so, sometimes whorled, broadly ovate, base cordate, margin entire or sinuolate, usually 3-lobed, apex acuminate, palmately 5-7-veined basally.

Flower: inflorescence a paniculate, terminal, Calyx bilabiate, corolla campanulate, pale yellow, purple spotted at throat.

Fruit: capsule, linear, nodding.

Distribution: from south of Liaoning to north of Guangdong, and many areas of southwest China.

* *Catalpa bungei* C. A. Mey. 楸树 is also planted commonly, but the leaf blade is triangular-ovate or ovate-oblong.

4.54 Verbenaceae 马鞭草科

Leaves opposite or rarely whorled, exstipulate, simple or compound. Flower zygomorphic or rarely actinomorphic. Calyx persistent. Ovary superior, ovules 1 or 2 per locule; style terminal. Fruit a drupe or indehiscent capsule, sometimes breaking up into nutlets.

There are 20 genera and 182 species in China.

1. *Vitex* Linn. 牡荆属

Leaves opposite, palmately 3-8-foliolate. Drupe subtended by enlarged calyx, globose to obovoid.

Vitex negundo Linn. 黄荆 (Figure 4-188)

Habit: shrubs or small trees, deciduous.

Bark: grayish brown.

Twig: 4-angled, densely gray tomentose.

Leaf: palmately, 3-7-foliolate, central leaflet distinctly petiolulate, margin entire or few serrates.

Flower: calyx campanulate, 5-dentate, gray tomentose; corolla 2-lipped, 5-lobed, outside puberulent.

Fruit: drupe, globose or ovoid.

Distribution: south provinces of the Yangtze River.

* *Vitex negundo* var. *cannabifolia* (Sieb. et Zucc.) Hand.-Mazz. 牡荆 is similar to the species, but the former leaflets distinctly serrate, abaxially sparsely pubescent.

2. *Clerodendrum* Linn. 大青属

Shrubs deciduous. Leaves simple, opposite or rarely whorled. Corolla with a slender tube, 5-lobed, spreading. Fruit a drupe.

Figure 4-188 *Vitex negundo* 黄荆

Clerodendrum cyrtophyllum Turcz. 大青（Figure 4-189）

Habit: shrubs or small trees.

Twig: yellow-brown, pubescent; pith white.

Leaf: Simple, oblong to elliptic, abaxially glandular, base rounded to cuneate, margin entire or rounded serrate, apex acuminate to acute, veins 6-10 pairs.

Flower: small, fragrant; calyx conspicuous, turn red when fruited; corolla white, tube slender, sparsely puberulent, lobes ovate. Inflorescence a corymbose.

Fruit: drupe, blue-purple, obovate to globose.

Distribution: east China, central China and southwest China.

Figure 4-189 *Clerodendrum cyrtophyllum* 大青

3. *Callicarpa* Linn. 紫珠属

Shrubs. Twig terete or 4-angled with hairs stellate or mealy tomentose. Leaves opposite, abaxially glandular. Fruit a small globose purple drupe.

Callicarpa bodinieri H. Léveillé 紫珠 (Figure 4-190)

Habit: shrubs; twigs, petioles, and inflorescences with dense stellate hairs.

Leaf: red glandular, adaxially pubescent; base cuneate, margin small serrate, apex acuminate to acute

Flower: pedicel short; calyx small, teeth obtusely triangular, stellate tomentose; corolla purple, stellate tomentose. Inflorescence a cymes, peduncle short, bracts small.

Fruit: purple, globose, very small.

Distribution: from central China, south China to southwest China.

* The species is similar to *Callicarpa cathayana* H. T. Chang 华紫珠 and *C. giraldii* Hesse ex Rehd. 老鸦糊, but *C. cathayana* without hair nearly, *C. giraldii* abaxially stellate hairs and yellow glandular.

Figure 4-190 *Callicarpa bodinieri* 紫珠

4.55 Lythraceae 千屈菜科

Branchelets often quadrangular. Leaves opposite, often decussate, or whorled, rarely subalternate to alternate, simple, entire. Inflorescence a racemes, cyme, or panicle; flower bisexual, regular or irregular. Ovary superior. Fruit a capsule with persistent floral tube, dehiscent.

There are 10 genera and 43 species in China.

Lagerstroemia Linn. 紫薇属

Trees or shrubs, deciduous. Bark glabrous. Young stems terete to frequently 4 - angled, glabrous. Petals 5-9, rose, purple, or white, crinkled, slenderly clawed. Fruit an indurate capsule, with persistent floral tube, loculicidally dehiscent, 3-6-valved.

Lagerstroemia indica Linn. 紫薇 (Figure 4-191)

Habit: shrubs or small trees, deciduous.

Bark: brownish yellow, glabrous.

Twig: slender, 4-angled or subalate, puberulous, glabrescent.

Leaf: sessile or with short petiole, elliptic, glabrous, base broadly cuneate to rounded, apex acute, obtuse with small mucro, or retuse.

Flower: inflorescence a panicle. Petals 6, purple, fuchsia, pink, or white, orbicular, margin crinkled.

Fruit: capsule, ellipsoidal.

Distribution: cultivated in many areas of China.

* The species is similar to *Lagerstroemia subcostata* Koehne 南紫薇, but the latter branchlet terete; leaf blade oblong or lanceolate, apex acuminate; capsule small, ca. 3-5mm diameter.

Figure 4-191 *Lagerstroemia indica* 紫薇

4.56 Scrophulariaceae 玄参科

Flower perfect, usually zygomorphic. Calyx often persistent. Corolla sympetalous; limb (3 or)4- or 5-lobed, often 2-lipped. Fruit a capsule, septicidal, loculicidal, or septifragal, sometimes opening by pores or irregularly dehiscent, rarely a berry.

There are 61 genera and 681 species in China, especially in southwest China.

Paulownia Sieb. et Zucc. 泡桐属

Trees, deciduous. Leaves opposite. Inflorescence a large pyramidal to cylindric thyrse. Flower large. Calyx leathery. Corolla purple or white, funnelform – campanulate to tubular –

funnelform; tube base constricted and slightly curved; limb 2-lipped; lower lip elongated and 3-lobed; upper lip yellow, 2-lobed. Capsule loculicidal. Seeds small, membranous winged.

Paulownia tomentosa **(Thunb.) Steud. 毛泡桐（Figure 4-192）**

　　Habit: trees, deciduous.

　　Bark: brown-gray.

　　Twig: conspicuously lenticellate, viscid glandular when young.

　　Leaf: cordate, abaxially densely to sparsely hairy, adaxially sparsely hairy, apex acute.

　　Flower: inflorescence a thyrse pyramidal to narrowly conical. Calyx shallowly campanulate, outside tomentose. Corolla purple, funnelform-campanulate, ridged ventrally, outside glandular, inside glabrous, with black spots and yellow stripes.

　　Fruit: capsule, ovoid, densely viscid-glandular hairy.

　　Distribution: cultivated or wild in China, especially in north China.

　　* The species is similar to *Paulownia fortunei* (Seem.) Hemsl. 白花泡桐, but the latter corolla white, purple, or light purple, ventral plaits inconspicuous; capsule oblong to oblong-ellipsoid.

Figure 4-192 *Paulownia tomentosa* 毛泡桐

4.57 Arecaceae 棕榈科

Mainly trees with stout unbranched stem ending in crown of leaves. Leaves large, alternate, young leaves plicate, exstipulate with long petioles. Inflorescence enclosed in a persistent spathe. Flowers unisexual; perianth 6 in two whorls of 3 each; in male flower 6 stamens in two whorls, anthers versatile; in female flowers carpels three. Fruit a berry or drupe; seed endospermic.

There are 18 genera and 77 species in China.

Trachycarpus H. Wendl. 棕榈属

Leaves palmate with long petiole, Inflorescence branched to 4 orders, covered with many sheathing bracts, usually yellowish at flowering time. Fruit a drupe, yellowish brown to purple-black.

Trachycarpus fortunei (Hook.) H. Wendl. 棕榈 (Figure 4-193)

Habit: trees, evergreen.

Trunk: shaggy and hairy with leftover dried petioles from shed leaves.

Leaf: leaf sheath fibers persistent, petiole long, margins with very fine teeth, blades semicircular in outline.

Flower: inflorescence erect. Male flowers yellow, female flowers greenish.

Fruit: drupe, yellow to blue-black.

Distribution: south provinces of the Qinling Mountains.

Figure 4-193 *Trachycarpus fortunei* 棕榈

4.58　Poaceae 禾本科[①]

Culms woody, perennial. Lower culm sheaths broad with rudimentary blades, nutrient leaf with parallel veins. Internodal regions of the stem hollow usually, vascular bundles in the cross-section scattered throughout the stem. Fruit a caryopsis.

There are ca. 37 genera and 500 species in China.

Phyllostachys Sieb. et Zucc. 刚竹属

Rhizomes leptomorph. Culms diffuse; internodes profoundly flattened or grooved on one side above branches; nodes 2-ridged. Branches 2, subequal.

Phyllostachys edulis (Carr.) J. Houzeau 毛竹 (Figure 4-194)

Habit: bamboo, evergreen, 20 m high or more.

Clum: internodes conspicuous; initially white powdery, densely puberulent

Figure 4-194　*Phyllostachys edulis* 毛竹

[①] In this family, we just focused on Sub Fam. Bambusoideae 竹亚科.

Leaf: culm sheaths yellow-brown with dark brown spots, densely brown hairy; leaf blade 4-11 cm, lanceolate to linear.

Flower: inconspicuous.

Fruit: caryopsis, narrowly elliptic.

Distribution: south provinces of the Yangtze River basin.

Phyllostachys nigra **(Lodd. ex Lindl.) Munro 紫竹 (Figure 4-195)**

Habit: bamboo, evergreen.

Clum: internodes green or gradually developing purple-brown to black spots or turning uniform purple-brown or black, initially white powdery, densely puberulent.

Leaf: culm sheaths red-brown, sheath ligule strongly convex, peaked or arcuate.

Flower: inconspicuous.

Fruit: caryopsis.

Distribution: native to Hunan, also widely cultivated in China.

Figure 4-195 *Phyllostachys nigra* 紫竹

References:

Bell A D, Bryan A, 2008. Plant form: an illustrated guide to flowering plant morphology[M]. Portland: Timber Press.

Alan F. Mitchell, 1978. A field guide to the trees of Britain and northern Europe[M]. Glasgow: Collins.

Press B, 2016. Green Guide to Trees of Britain and Europe[M]. London: Bloomsbury Publishing

Nelson G, Earle C J, Spellenberg R, 2014. Trees of Eastern North America[M]. Princeton: Princeton University Press.

Simpson M G, 2019. Plant systematics[M]. Burlington: Academic press.

吴国芳, 1982. 植物学(下册)[M]. 北京: 高等教育出版社.

汪劲武, 2009. 种子植物分类学[M]. 北京: 高等教育出版社.

祁承经, 汤庚国, 2005. 树木学: 南方本[M]. 2版. 北京: 中国林业出版社.

张志翔, 2008. 树木学: 北方本[M]. 北京: 中国林业出版社.

江苏植物研究所, 1977. 江苏植物志(上册)[M]. 南京: 江苏人民出版社.

江苏植物研究所, 1982. 江苏植物志(下册)[M]. 南京: 江苏科学技术出版社.